THE ATLAS OF
THE WORLD'S MOST
DANGEROUS
ANIMALS

THE ATLAS OF THE WORLD'S MOST DANGEROUS ANIMALS

MAPPING NATURE'S BORN KILLERS

PAULA HAMMOND

Marshall Cavendish
Reference

NEW YORK

This edition first published in 2010 in the United States of America
by Marshall Cavendish.

Marshall Cavendish
99 White Plains Road
Tarrytown, New York 10591-9001

www.marshallcavendish.us

Library of Congress Cataloging-in-Publication Data

Hammond, Paula.
  The atlas of the world's most dangerous animals : mapping nature's born killers / Paula
Hammond.
     p. cm.
  Includes index.
  ISBN 978-0-7614-7870-6
  1. Dangerous animals. 2. Dangerous animals--Geographical distribution. I. Title.
  QL100.H335 2010
  591.6'5--dc22

                         2008044960

Printed in China

13 12 11 10 09 1 2 3 4 5

Editorial and design by
Amber Books Ltd
Bradley's Close
74–77 White Lion Street
London N1 9PF
United Kingdom
www.amberbooks.co.uk

Project editor: Tom Broder
Designer: Joe Conneally

Artwork credits: International Masters Publishers Ltd

Picture credits:
Corbis: 10, 15. 26, 117, 185
NHPA: 19, 72
Natural Science Photo: 23 (Carlo Dani & Ingrid Jeske), 34 (P.H. & S.L. Ward), 39 (D.
Allen Photography), 42 (Jim Merli), 49 (P.H. & S.L. Ward), 56 (Steve Downer), 60 (Carlo
Dani & Ingrid Jeske), 69 (J. Hobday), 78 (C. Banks), 82 (C. Mattison), 86 (Carlo Dani &
Ingrid Jeske), 90 (Beth Davidow), 95 (Carlo Dani & Ingrid Jeske), 102 (D. Yendall), 112
(Jim Merli), 121 (Ken Cole),132 (Hal Beral), 139 (Jim Merli), 146 (Brian Gibbs), 150
(Mary Clay), 154 (Carlo Dani & Ingrid Jeske), 158 (Pete Oxford), 164 (Paul Hobson),
168 (Richard Revels), 173 (R.P.B. Erasmus), 177 (Richard Revels), 180 (G. Kinns), 194
(Ken Hoppen),199 (Ken Hoppen), 202 (Bob Cranston), 206 (Kjell Sandved), 214 (C.
Banks), 219 (Hal Beral)

Contents

Introduction

Siberian Tiger

Lammergeier

To humans, the most dangerous
animals would seem to be
those that put us at most
risk — the man-killers.
Sharks, big cats,
bears and snakes: every
continent has its own candidate
for this gruesome Hall of Fame. From the
Asian King Cobra to the North American Grizzly, these fearsome
predators kill and maim hundreds of thousands of people every year.
Yet not all killers are meat-eaters. Rogue hippopotami and stampeding
elephants are just as dangerous and, in many cases, responsible for
more human deaths than the natural-born killers. On a worldscale,
though, tigers and enraged hippos are mere beginners. More human
deaths are caused every year by insects like the
humble locust, which devastate crops and
bring wide-scale famine.

Wolf Spider

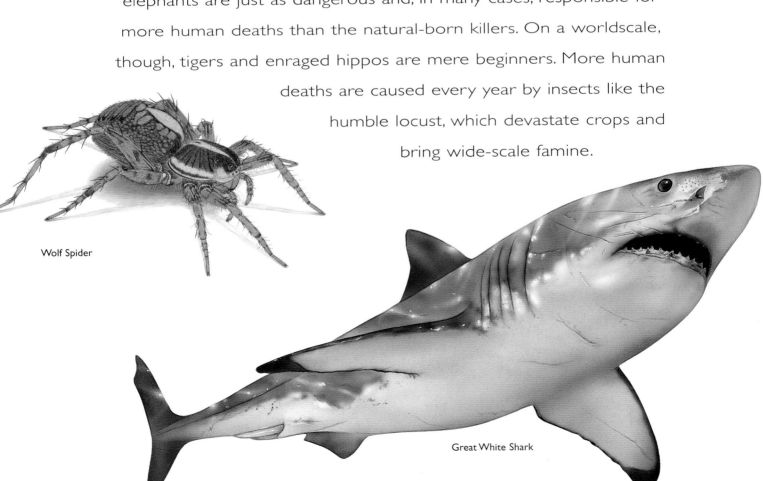

Great White Shark

Green Tree Python

Nile Crocodile

To a Thomson's Gazelle, the most
dangerous animal is undoubtedly the cheetah, who
lists this lithe member of the antelope family amongst its
favourite food. To the snake, it's probably the mongoose,
that fast, tenacious little mammal who specializes in
making meals out of one of nature's most feared killers.
And for members of the insect family, the Wolf Spider seems
just as deadly as its full-sized namesake. Whether an animal is
dangerous, depends on your point of view. The English
poet Alfred Tennyson (1809–1892) rightly said that
nature was 'red in tooth and claw', and in the animal
kingdom the truly dangerous ones are those who want to
put you on the menu.
In this volume, we hope to explore the world of
dangerous animals from all these perspectives, giving
you a fresh look at the man-killers, the predators,
the great hunters, and the perhaps
not so gentle herbivores.

Strawberry Poison-Arrow Frog

Frilled Lizard

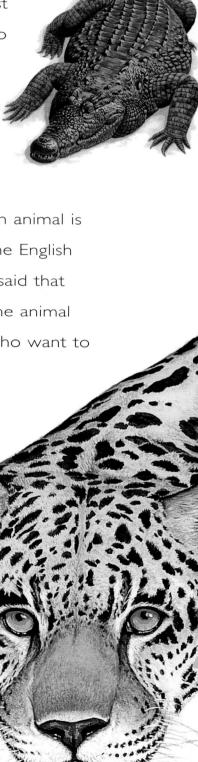

Jaguar

EUROPE

MEDITERRANEAN SEA

Atlas Mountains

Libyan
Desert

Nile

Nubian
Desert

Sahara

AFRICA

ARABIA
SEA

Congo
Basin

Rift Valley

Lake
Victoria

Lake
Tanganyika

INDIAN
OCEAN

Lake
Nyasa

MADAGASCAR

Namib Desert

Kalahari
Desert

SOUTH ATLANTIC
OCEAN

Cape of Good Hope

Africa

Africa is a continent of contrasts. From lush, green tropical rainforests to harsh desert sands, nowhere on Earth can we find such astounding variety and natural beauty.

The second largest of the world's seven continents, Africa covers 300,330,000 square kilometres (11,700 square miles). Yet, if we were to view this immense landmass from space, the first thing we'd notice is that most of it is comprised of a huge, wide plateau.

To the north is the great Sahara Desert, which covers an area larger than the entire United States. On the edges of this vast ocean of sand, desert slowly merges with grasslands to produce swaths of golden, sun-scorched savannah. Snaking almost the length of the whole continent is the mighty Nile, the world's longest river, which brings water and life to the parched interior. In Central Africa are the continent's great rainforests, a startling carpet of green nestled in the vast Congo basin. Towards the coast, the picture is completed by narrow strips of bustling coastline.

Such a variety of habitats has made Africa home to some of the world's most famous and spectacular wild animals – plus a few surprising ones too. On the South African coast, for example, we can find penguins, who seem much more comfortable basking on tourist-filled beaches than in the frozen Antarctic. It's in the thick dense jungles, steamy swamps and rolling plains, though, that we find Africa's most familiar and dangerous inhabitants and where, every day, hunter and hunted play out their deadly game of survival.

African Elephant

Over 1.6 million years, the African Elephant has evolved into the world's largest land mammal. When roused, these seemingly gentle giants can charge at up to 40km/h (25mph) in a terrifying and unstoppable stampede that brings death and destruction in its wake.

Key Facts

	ORDER *Proboscidea* FAMILY *Elephantidae* GENUS & SPECIES *Loxodonta africana*
Weight	Male up to 6000kg (13,228lb)
Length	Male up to 4m (13ft), head to rump female up to 3.3m (10ft 10in)
Shoulder Height	Male up to 3.27m (10ft 9in)
Sexual maturity	About 10 years
Breeding Season	All year
Gestation period	22 months
Number of young	1
Birth Interval	3–4 years
Typical diet	Grasses, foliage, shrubs, fruit, flowers, roots
Life Span	50–60 years

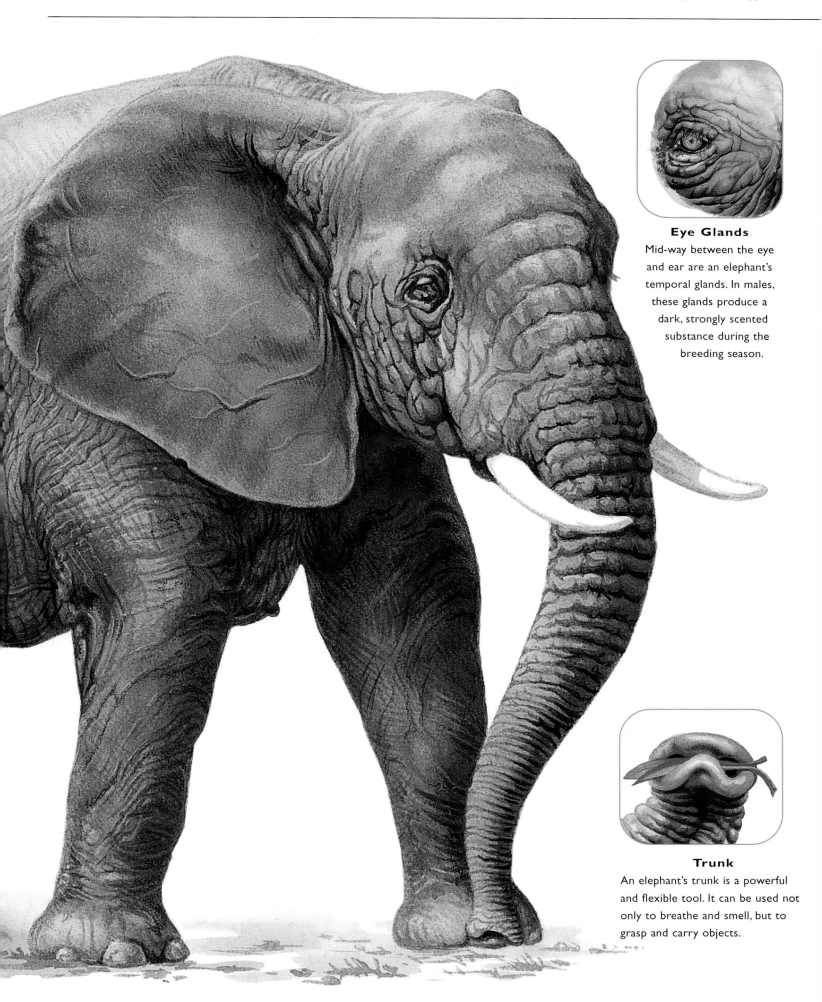

Eye Glands
Mid-way between the eye and ear are an elephant's temporal glands. In males, these glands produce a dark, strongly scented substance during the breeding season.

Trunk
An elephant's trunk is a powerful and flexible tool. It can be used not only to breathe and smell, but to grasp and carry objects.

Comparisons

As these pictures show, there are a number of clear differences between African and Indian Elephants. Indian Elephants are generally smaller, with lighter skin and less prominent tusks. They also tend to have a slightly humped back and two lumps on their forehead. Up close, there are less obvious differences. An Indian Elephant, for example, has five toes on its front feet and four on the back. An African Elephant has four or five toes on its front feet, but only three at the back. It is the ears that are the biggest giveaway: the ears of African Elephants are larger and are shaped, some people say, like the African continent.

African Elephant

Indian Elephant

Modern elephants are the last surviving relatives of the Woolly Mammoth, which became extinct around 4000 years ago. Mammoths belonged to a group known as Proboscideans, who, like elephants, had elongated snouts or trunks. Once common throughout Europe, Asia, Africa and the Americas, elephants have been rapidly declining in numbers since the 1970s. African Elephants are larger than their Asian relatives, but there are now only about 500,000 remaining. These include the Forest Elephant of Central Africa and the West African Elephant, which lives in both forests and savannahs, but it is the Savannah Elephant, living south of the Sahara Desert, that is largest and the undisputed king of the grasslands.

Family Ties

Wild elephants live for around 60 years. During this time, they form close-knit family groups of about 10 or 12, headed by a dominant, older female called the matriarch. Elephants are generally social animals. They enjoy the company of the herd and are extremely expressive and communicative, using a series of low stomach rumbles as well as touch, scent and body posture to 'bond' with the rest of the group.

Male (bull) elephants generally stay with the herd only until they're about 12 years old, when they leave to form troops of their own. Exiling mature males from the family group in this way may be a safety measure. An adult bull elephant has glands between the eye and ear, which become active for about three months each year. During this time, the mature males enter a condition called must, which means madness. In this heightened state of sexual arousal, these elephants are very dangerous and frequently use their tusks and huge bulk to gouge and trample other elephants and, occasionally, humans.

Ancient Tanks

In the second century BC, the great North African general Hannibal (247–183 BC) crossed the Alps into Italy and

African Elephant habitat

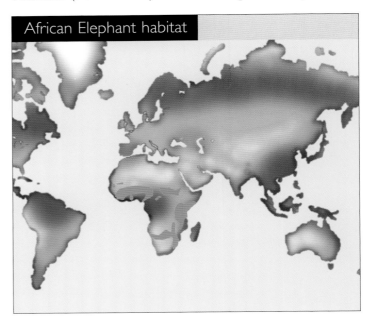

declared war on the Roman Empire. He took with him 26,000 soldiers, 6000 horses and dozens of war elephants. Elephants were the ancient world's equivalent of the tank. On a battlefield, they caused panic, not only because of their great size, but owing to their unpredictable nature. They could easily smash apart an enemy army, but were just as likely to inflict damage on their own troops. These enormous mammals have poor eyesight, so when alarmed or threatened they will charge blindly towards the source of the noise. Bull elephants can weigh up to 6 tonnes (6.5 tons), with tusks growing to 3m (10ft) in length, so there's little that can stop an angry or distressed elephant from stampeding. Hannibal's elephant riders were aware of this, and came equipped with a hammer and a huge, metal spike. If at any time they lost control of the elephant, the only way to stop it was to drive the spike into its brain.

Stampede!

It's in extreme conditions that elephants are at their most dangerous, yet not all elephant attacks are obviously defensive. Elephants are extremely intelligent, complex animals and sometimes their behaviour seems almost human. In 2002, for example, a herd of elephants attacked a village that was encroaching on their range. After drinking stores of beer, they went on what seemed like a drunken rampage, pulling down homes, destroying crops and killing whoever stood in their way.

In another village, an elephant attacked a man who was trying to hide from the herd up a tree. The elephant shook him down from the tree and used his powerful, column-like legs to trample him to death. Later, he bathed the body and stood guard over it, almost as though he regretted his actions.

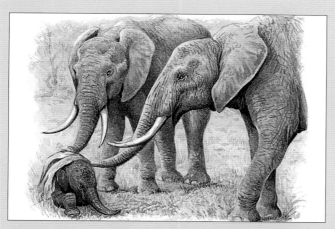

From the moment of birth, a newborn becomes the centre of attention within the family group.

If necessary, the young calf is gently helped to its feet, either by the mother or by another female.

The mother keeps a close watch over the calf on the move, ready to steady the newborn with her trunk should it falter.

The mother shields the calf from the fierce sun during the heat of the day.

Cheetah

Able to accelerate up to 96km/h (60mph) in just three seconds, the cheetah is the world's fastest land animal. For this big cat, however, hunting is all about skill and timing. If it doesn't catch its prey within 30 seconds, it will be too exhausted to continue the chase.

Forelegs
Cheetahs are able to stretch their legs much further than other animals. This allows them to cover vast distances when running.

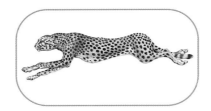

Spine
Shoulder blades lie to the side of the body, so cheetahs can arch their spines to cover distances at speed.

Key Facts

Key Facts	ORDER *Primates* / FAMILY *Pongidae* GENUS & SPECIES *Pongo pygmaeus*
Weight	34–68kg (75–150lb)
Length Head and body Tail	1.3–1.5m (4ft 3in–4ft 11in) 60–80cm (24–31in)
Sexual maturity	20–24 months
Mating season	Throughout the year
Gestation period	90–95 days
Number of young	Up to 8, but usually 2 to 5
Birth interval	17–20 months
Typical diet	Gazelles and other antelope species; also hares, rodents and gamebirds
Life Span	Up to 12 years in the wild

Claws

Most cats pull back their claws when not in use. Cheetahs can't, using them instead for grip while running.

Grasslands and Cheetah habitat

The cheetah can be found across regions of grassland and semi-desert throughout Africa. These flat, open areas allow the cheetah to make use of its excellent vision and great speed.

Desert

Grasslands

Semi-desert

Coniferous forest

Tropical forest

Evergreen woodland

Range of the cheetah

Cheetahs once inhabited the grassy plains of Africa, the Middle East and Central Asia. Here they were sometimes kept as pets by wealthy landowners, who called them 'hunting leopards'. Perhaps because they're so beautiful and graceful, the perception that these powerful and skilled killers can be tamed persists to this day. Some cheetahs can even be found on ranches where tourists are encouraged to pet them like house cats, with the occasional, and predictable, bloody result. In the wild, however, cheetahs are typically found in wide, open savannahs and on scrub and bush land, where they're increasingly in danger from human encroachment and hunters.

Designed To Perfection

Cheetahs are the ultimate 'designer predator'. Every inch of this unique cat – from its small round head to its rudder-like tail – has been shaped by nature over thousands of years for maximum speed and agility. Their bodies are naturally slender and streamlined. They have extremely long, muscular legs and a flexible spine. This gives them an immense 7m (23ft) stride, which enables them to cover huge distances at a single bound. Despite being members of the family *Felidae* (cat family), cheetahs also have a number of doglike qualities. They can't, for example, climb trees. Nor can they retract (pull back) their claws into their paws

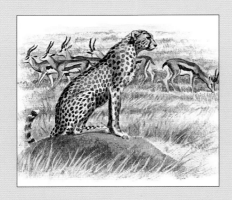

A cheetah watches a herd of gazelle from a vantage point nearby, trying to identify a possible quarry.

The cheetah closes to within 100m (328ft) before launching into a sprint. Long strides and incredible agility bring it closer and closer to its prey.

when these are not in use. Instead they have blunt, round-ended claws, like a dog, which give them added traction on the ground as they run. Finally, they have excellent camouflage. Their coats are cream-coloured with solid, circular black spots. So, by crouching in the tall savannah grasses, they can get within 30m (100ft) of their prey without it even noticing. All these adaptations make them amazingly efficient hunters.

Swift and Deadly

Unlike lions, cheetahs are solitary predators who mainly hunt by day. Their natural prey are small, nimble mammals such as the springbok or impala. Their favourite, though, is the Thomson's Gazelle and 80 per cent of all cheetah kills are from this species.

Once a cheetah chooses a victim, it locks its eyes on its quarry and follows its every move. During the chase, the cheetah's body turns, twists and changes direction to match the movements of its prey exactly, in an amazing demonstration of speed and agility. Cheetahs expend so much energy in this athletic display that, within half a minute, their body temperatures rise to almost fatal levels. Gazelles and antelopes may not be as fast as cheetahs, but they do have more stamina. So, the secret of survival for the lowly grassland bovine is simply to keep running until the cheetah gets tired!

Combined Success

Perhaps because hunting can be so strenuous – and so hit and miss – cheetahs' social structures are amazingly flexible. Females tend to live alone or with their cubs in small territories that may overlap those of other females. Males are usually loners, living a roaming, nomadic lifestyle over ranges of around 800–1500 square kilometres (310–580 square miles). Yet some occasionally live in groups with up to four other males. These cooperatives last for the cheetah's whole life, which is some 12–14 years in the wild. Even in a cooperative, cheetahs may still hunt separately, but working together increases their chances of a successful kill. Groups of cheetahs are also able to deal with much larger prey: it has been known for just two to bring down an animal as big as a wildebeest (gnu), which can be at least six times as heavy as the average cheetah. That's like a human being wrestling a Polar Bear and winning!

Comparisons

A cheetah's spots form part of its natural camouflage. It may seem that these startlingly black splashes of colour would make a cheetah more, not less visible. Yet, other big cats use similar bold camouflage patterns. The reason is simple: it works! These dashes of black break up the cheetahs outline, making it hard for prey or other predators to see them clearly. As can be seen in the pictures below, these camouflage patterns vary between cat species. The spots on a leopard and jaguar form a rosette pattern, while the cheetah's spots are more open and regular in shape.

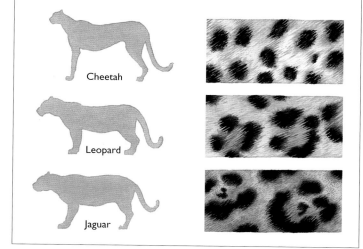

Cheetah

Leopard

Jaguar

When within striking range, the cheetah lashes out at the hind quarters of the prey. The legs are knocked from beneath the fleeing animal.

The quarry falls. In that moment, the cheetah clamps its vicelike jaws around the gazelle's throat, suffocating the unfortunate animal.

Fat–Tailed Scorpion

Typically no longer than 12cm (4in), the Fat-Tailed Scorpion has a formidable reputation. A sting from this small armoured hunter can kill a man in seven hours, and a dog in just seven minutes.

Jaws

Using a scissor-style motion, powerful jaws are used to crush and shred prey. The victim's bodily fluids are then sucked up and digested.

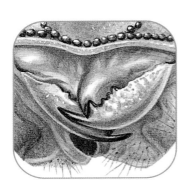

Key Facts	ORDER *Scorpiones* / FAMILY *Buthidae* GENUS & SPECIES *Androctonus australis*
Weight	25g (1oz)
Length	10cm (4in)
Sexual maturity	6 months to 3 years
Mating season	Unknown
Gestation period	About 6 months
Number of young	40 to 50
Breeding interval	1 year
Typical diet,	Invertebrates, such as beetles, cockroaches and spiders
Lifespan	3–5 years

Tail and stinger

Just before a scorpion attacks, it will arch its tail over its body, ready to strike. At the very tip of this tail is the stinger, which is used to inject venom into its victim.

Eyes

Scorpions have a pair of eyes in the centre of their head, plus a further group of three to either side of the head. Despite this, they have poor eyesight.

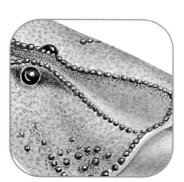

Scorpions have been found in fossils dating back some 400 million years. This makes scorpions one of the world's most ancient arachnids. They are at home in the tropics, but species have nevertheless been found as far north as British Columbia. The Fat-Tailed Scorpion is especially widespread, and can be found in Egypt, Somalia, Saudi Arabia, Israel and Pakistan.

Family Trees

One of the most accepted ways of classifying animals is to group them according to their biological ancestry. Under this system, every animal belongs to a kingdom, a phylum, a class, an order, a family, a genus and a species. For example, a tiger is an animal, so it belongs to the kingdom *Animalia*. It has a backbone, which makes it part of the phylum *Chordata*. It's also a mammal, which puts it in the class *Mammalia*. Finally, tigers are carnivores (order *Carnivora*) and members of the cat (*Felidae*) family, of the genus *Panthera*.

Scorpions belong to the phylum *Arthropoda* (anthropoid), class *Arachnida* (arachnid), order *Scorpiones* (Scorpion). Like all arachnids, scorpions are small, with a body divided into two main parts. The front part, called the cephalothorax, includes the head and thorax (chest). The hind part is called the abdomen. Being an arachnid, not an insect, scorpions also have also eight legs. Despite being small, arachnids are dangerous predators and the scorpion is no exception. Here, offensive weaponry comes in the shape of a pair of powerful, crushing claws and a long, segmented tail that contains potent poison.

Fat-Tailed Scorpion habitats

Deadly Toxin

Although we're most familiar with the concept of venomous reptiles, especially snakes, almost every class of animals throughout the animal kingdom uses venom. This includes some unlikely suspects, like the Duckbill Platypus, some species of starfish and even a few snails. Only birds (of the order *Aves*) are the exception.

Practically all arachnids carry some form of venom, and many scorpions have a particularly potent blend. There are about 1500 species of scorpions in the world and an estimated 25 of these are capable of killing a human — in Mexico and the USA, more people are killed each year by

Comparisons

As most scorpions live in desert regions, they commonly come in shades of brown or sandy yellow, which provides excellent natural camouflage. The exception is the almost jet black Emperor Scorpion, which is one of the largest scorpions, at 20cm (8in) long. At just 1cm (^{3}/8in), the tiny *Microbuthus pusillus* is so small that protective colouration hardly matters!

Fat-Tailed Scorpion

Microbuthus pusillus

Emperor Scorpion

scorpions than by snakes. It is believed that the Fat-Tailed Scorpion kills around 400 people a year in Tunisia alone. In fact, its venom is so dangerous that it has been compared in strength to the bite of a King Cobra.

Defence and Offence

A scorpion's venom is held in two glands below the stinger, which is a sharp point at the tip of the tail. Just before a scorpion attacks, it arches its tail over its body. Holding its prey in its claws, the scorpion then injects its poison. A typical feast for a Fat-Tailed Scorpion might be a spider, beetle or insect, but being one of the smaller species of scorpion doesn't prevent the Fat-Tailed Scorpion from having big ideas. One of the major advantages of venom is that it allows even modest-sized predators to kill much larger prey. So the Fat-Tailed

Scorpion also eats the occasional small mammal and rodent. Once a victim has been poisoned, the scorpion then dismembers the body, using its razor-sharp jaws to cut the flesh into pieces small enough to eat.

Scorpion venom is a very complex mix of toxins and, strangely, a Fat-Tailed Scorpion's poison seems to have been specially designed to kill vertebrates. As most of its prey comprises invertebrates, this would seem to suggest that scorpion venom was originally developed for protection rather than hunting. As if to emphasize this, a scorpion's venom contains chemicals specially designed to cause pain. So a scorpion may not be able to kill an enemy, but it can at least make it think twice about continuing an attack. Scorpions can also spray their venom up to 90cm (3ft). This can cause temporary blindness – a very effective way to deter aggressors.

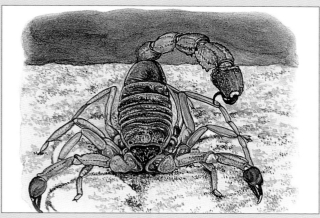

The concealed scorpion waits for its prey. Vibrations alert the scorpion when prey approaches.

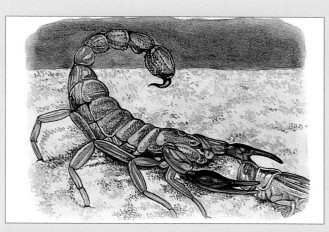

An unfortunate locust passes too close. The scorpion pounces and holds the locust in its powerful pincers.

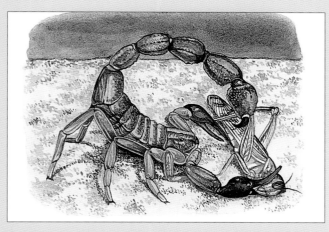

The scorpion arches its tail over its body to deliver a lethal dose of venom into the struggling victim.

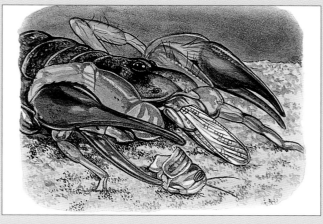

Although the locust is shredded by the scorpion's jaws, the scorpion is unable to digest the outer body parts, and so ingests the bodily juices of its prey instead.

Chimpanzee

Of all the great apes, it is the chimpanzee, that most human-seeming of animals, that fascinates us. Like us, chimpanzees are capable of great tenderness and consideration to other members of their troop, but they can also be predatory, aggressive and extremely violent.

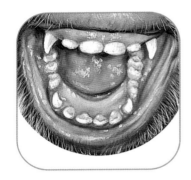

Teeth

As chimpanzees have a varied diet, their teeth – like ours – have developed to cope with a range of different food-stuffs.

Feet

Chimpanzee's feet have long, dextrous toes – very much like a 'spare' pair of hands. These can be used to grip objects, such as tree branches.

Hands

The secret of the chimpanzee's success is its 'opposable' thumb. Like humans, chimps are able to use their thumb and fingers in combination to manipulate objects with skill.

Key Facts

ORDER *Primates* / FAMILY *Pongidae*
GENUS & SPECIES *Pan troglodytes*

Weight	27–68kg (60–150lb)
Length	71–91cm (28–36in)
Sexual maturity	7 years, but does not breed until 12–15 years old
Mating season	All year
Gestation period	230 days
Number of young	I
Birth Interval	3 years, often longer
Typical diet	Fruit, leaves, berries, nuts, insects and, occasionally, mammals and birds
Life Span	Up to 60 years

Chimpanzees, gibbons, gorillas and orang-utans are all anthropoid apes of the family *Pongidae*. This means that they share many characteristics linking them to humans, but are distinguished from other primates: by having no tails, walking with an upright posture and having a highly developed brain. Of this group, it is the chimpanzees who are the most complex and accomplished.

Jungle Life

Found in a band across Africa, from the Niger Basin to Angola, chimpanzees are most at home in the rainforests, where they're perfectly adapted for a life in the treetops. Being apes, not monkeys, they don't have tails, but instead use their strong, dextrous hands and prehensile feet to clamber from tree to tree. At night, chimps use this natural dexterity to make large nests from branches and foliage,

where they can sleep in relative safety from predators. During the day, chimpanzees often spend time on the ground in search of food. A fully grown male chimpanzee has a huge appetite and can consume 50 bananas in one sitting. It's not unusual for them to spend up to seven hours a day foraging. While on the ground, chimpanzees will often run on all fours, but they can walk upright if necessary – for example, to see over tall grass and look for danger. An enraged male will also often stand upright to demonstrate his dominance to the rest of the group. Screaming, waving branches and baring his teeth, an angry chimpanzee can be a terrifying and awe-inspiring sight.

Good Mothers, Bad Mothers

Naturally, chimpanzees, like humans, live in a variety of social groupings. Some chimps live in large troops with up

When the group forages for food, the chimpanzees travel in single file on all fours.

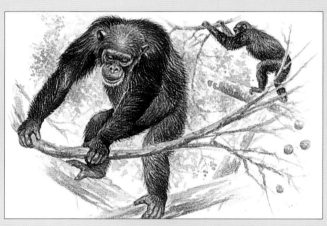

The group finds a tree with plenty of nuts. Two of their number climb up to shake the nuts from the branches.

Down on the ground below, the remaining chimpanzees collect the nuts as they fall.

Two young chimpanzees sit watching, learning from the adult as she uses a stone to break open the nuts.

to 40 other males, females and babies; some live in all male groups; while some chimpanzees are loners and live a solitary existence in the forest. The only constant members of the group tend to be mothers and babies. Adult females (those over the age of 12), mate between August and November and give birth to one baby at a time, which they raise as part of the group.

Unusually for the animal kingdom, chimp mothers are not always natural parents. It seems a very human trait, but some inexperienced mothers appear baffled about how best to care for their new charges. Others may spoil and coddle their youngsters to such an extent that it's not unusual to find chimps as old as seven still being fed by their doting mum.

Dangerous Rivals

The more we learn about chimpanzees, the more they surprise us. In the 1950s, the generally accepted view was that they were gentle vegetarians. Zoologists now know that their behaviour is much more complex. For example, a chimpanzee's diet is mostly made up of fruit or insects, but they do hunt other animals, including monkeys and other apes, for meat. Such hunts are planned with military–style precision: one chimp drives the enemy out of its cover, while three 'blockers' wait in ambush.

Meat is a valuable source of protein and many 'vegetarian' animals often supplement their diet in this way, but some chimps seem to have made an aesthetic choice: they prefer the taste of meat. What's more, chimpanzees have been known to make war on their neighbours, in a startlingly human display of aggression.

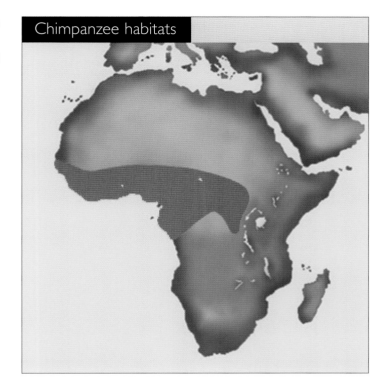

Chimpanzee habitats

Often, when such conflicts occur, males will kill and eat the other troop's babies, presumably in an attempt to reduce their rivals' numbers. In Uganda, chimpanzees have recently begun to attack human children in the same way. As the human population encroaches ever futher into the chimpanzees' natural habitats, it seems that they have started to recognize the fact that it's us – rather than other chimpanzee troops – who are their most dangerous rivals for food and land.

Comparisons

Although they are close relatives to the chimpanzee, gorillas are larger and much more powerful. On average, a fully grown male gorilla (called a silver back, because of its 'greying' fur) weighs around four times as much as a chimp.

Gorilla

Chimpanzee

Hippopotamus

In spite of their lumbering, genial appearance, hippopotami kill more people in Africa than any other animal. These formidable giants can weigh up to 4.5 tonnes (5 tons) and grow to 1.4m (4ft 7in). A yawning hippo may seem harmless, but this display of tusks and teeth is actually a warning to stay well away.

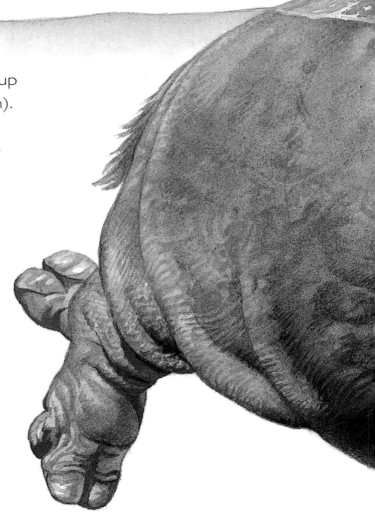

Key Facts	ORDER *Artiodactyla* FAMILY *Hippopotamidae* GENUS SPECIES *Hippopotamus amphibius*
Weight	Male 1000–4500kg (2205–9921lb); Female 1000–1700kg (2205–3748lb)
Length Head and Body Tail	2.9–5m (4ft 6in–16ft 5in) 40–56cm (15–22in)
Shoulder Height	1.5–1.65m (4ft 11in–5ft 5in)
Sexual maturity	Male 6–13 years; female 7–15 years
Mating season	Dry season, varies according to location
Gestation period	227–240 days
Number of young	1 (twins are rare)
Birth Interval	About 2 years
Typical diet	Mainly grass and aquatic vegetation
Life Span	35–50 years

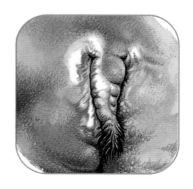

Tail
When marking their territory, male hippopotami spin their tails, to spray urine and faeces over the boundaries of their land.

Skin Glands

Hippos have very little hair to protect themselves from the sun. Instead, glands produce a sticky pink sun block!

Tusks and Teeth

A hippo's tusks are actually over-sized teeth. Those in the lower, outer part of the jaw are incisors. The large grinding teeth in the top jaw are molars.

Comparisons

As might be expected, the Nile Hippopotamus lives mainly along the southern most banks of the River Nile, which stretches from the Equator to the Mediterranean Sea. The Pygmy Hippopotamus prefers to make its home in rainforests and swamp lands. Being much smaller than their Nile cousins, pygmies are able to use the undergrowth effectively to hide from any potential danger. This means that they are less reliant on the water for protection, and so lack many of the natural aquatic adaptations which Nile hippos have developed.

Pygmy Hippopotamus

Nile Hippopotamus

Hippopotamus is Greek for 'river horse'. When submerged, the heads of these large, barrel-shaped mammals do look curiously horselike, yet they're actually more closely related to pigs. They were once numerous in rivers throughout Africa, but there are now only two known species of hippopotami: the Common, or Nile Hippopotamus, and the rarer, smaller Pygmy Hippopotamus. Both are found in Southern, Central and West Africa.

Hippo habitats

Big is Beautiful

Hippos in the wild live for 40–45 years, and much of this time is spent in or close to rivers and lakes. They even mate and give birth under water. When submerged, the water supports a hippo's huge bulk and enables it to move around remarkably gracefully. Like many animals, the hippopotamus has a body that has been adapted, over millions of years, to suit the needs of its particular lifestyle. Eyes, ears and nostrils at the top of their head enables them to see and hear perfectly, even when the bulk of the body is totally submerged. When diving, they can close their ears and nostrils to keep out the water, while their hooflike toes are webbed to help propel their hairless bodies forward. These adaptations have made hippopotami superbly versatile and a baby hippo is able to walk, run and swim within five minutes of being born.

Adaptable

Hippopotami are herbivores, which means that they're vegetarians. At night, hippos come onto dry land to graze on grass and eat fallen fruit. They're so efficient that, when they eat, they crop the grass so close to the ground that forest fires are unable to spread in the areas where they live. In one evening, an hippo can eat around 40kg (88lb) of grass. Surprisingly, this is actually quite a small amount for an animal of this size, so it would seem that their semi-aquatic lifestyle is energy-saving too. Being opportunists, hippopotami have also been known to eat small mammals or scavenge meat from discarded animal carcasses, especially in times of drought, when their natural food may be scarce.

Law and Order

Hippopotami are sociable animals and often gather in large family groups of 10 to 20 – sometimes even 100 – to share wallows (temporary watering holes). Yet, they're also extremely territorial, and to get along in hippo society, there are strict rules that must be obeyed. In the centre of each territory is a sandbank calld a crèche, used by females and their young. Around the crèche are areas known as refuges, occupied by males, who fight each other for access to the inner refuges, where they'll be able to mate with the females. All hippos have long, tusklike teeth and powerful jaws, so these fights can be violent, and sometimes fatal.

It's female hippopotami who can be the most dangerous, though, especially when defending their young. During the breeding season, males may visit the females in the crèche, but only if they show absolutely no signs of aggression. To demonstrate this, the male must lie down as soon as a female stands up. If he breaks this rule, he's considered as a threat to the safety of the crèche and will be driven off violently, as will any other animal (including humans) coming within range.

When hippos feel threatened, they usually head straight for water, where they're better able to defend themselves. However, they're by no means helpless on land. Although they may look ungainly and lumbering, an enraged hippo can run at 32km/h (20mph), which is fast enough to chase down a man or, over distance, a lion, which can manage only short spurts of speed.

Female hippos and their young congregate in a nursery group. They are protected by a dominant male, which means that they enjoy a relatively calm existence.

The male, however, is kept busy, patrolling around the nursery, on the lookout for any rivals who may dare to challenge his claim on the females.

He notices another male hippo trying to sneak into the nursery to mate. Defending his breeeding rights over the females, the dominant male swims to attack.

A violent battle ensues as the hippos use their tusks to slash and stab. Fights can be fatal.

Lion

Lions are the most famous of all of Africa's large carnivorous cats. With powerful, bear-like forelegs and curved claws that hook and hold their prey, a fully grown lion weighing 180kg (176lb) qualifies as one of Africa's most powerful predatory animals.

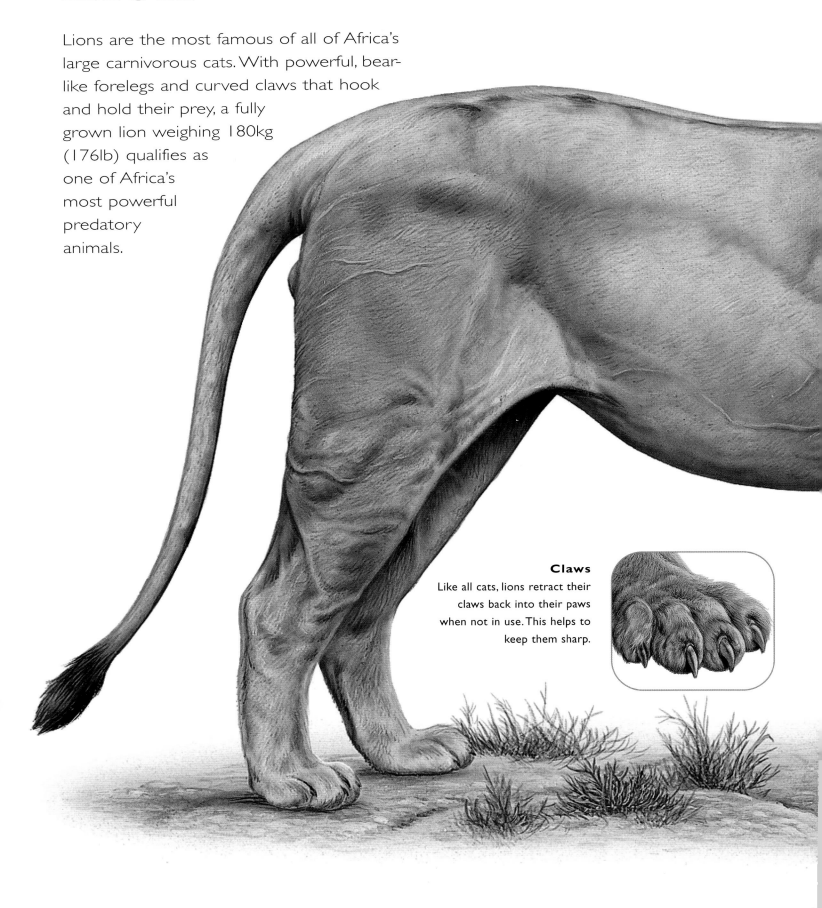

Claws

Like all cats, lions retract their claws back into their paws when not in use. This helps to keep them sharp.

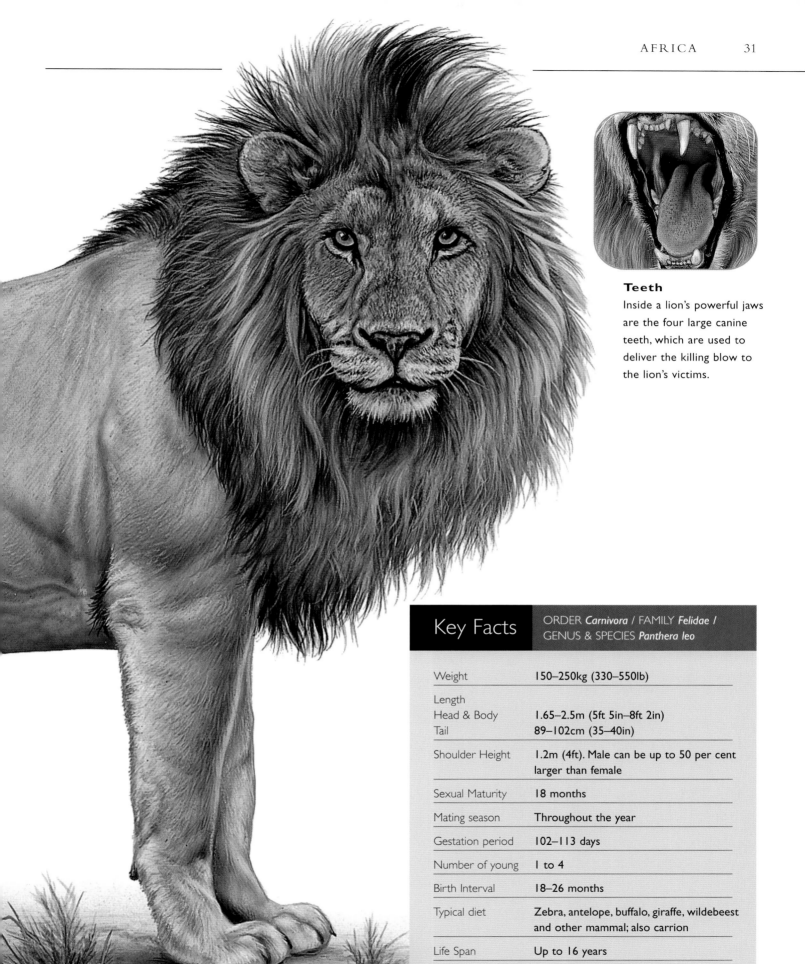

Teeth
Inside a lion's powerful jaws
are the four large canine
teeth, which are used to
deliver the killing blow to
the lion's victims.

Key Facts	ORDER *Carnivora* / FAMILY *Felidae* / GENUS & SPECIES *Panthera leo*
Weight	150–250kg (330–550lb)
Length Head & Body Tail	1.65–2.5m (5ft 5in–8ft 2in) 89–102cm (35–40in)
Shoulder Height	1.2m (4ft). Male can be up to 50 per cent larger than female
Sexual Maturity	18 months
Mating season	Throughout the year
Gestation period	102–113 days
Number of young	1 to 4
Birth Interval	18–26 months
Typical diet	Zebra, antelope, buffalo, giraffe, wildebeest and other mammal; also carrion
Life Span	Up to 16 years

Comparisons

Leopards are one of the most widespread of the Family *Felidae*, with sub-species found throughout Africa, as far North as the Savannah Desert, into the snow-swept mountains of China. Although only 60 percent of the size of a fully grown lion, the leopard is much faster and more agile than its larger African neighbour. Yet, unlike the lion, leopards tend to be solitary hunters, specializing in small mammals, like rabbits, wild sheep and goats.

Lion

Leopard

The lion is often referred to as the 'King of the Jungle', but these powerful cats don't actually live in jungle forests. They live where there's an abundance of food such as deer, zebra and buffalo, and enough scrub land and grasses to enable them to stalk their prey effectively. Although a few Asiatic Lions can still be found in the Gir Sanctuary in India, many sub-species, including the Persian Lion, are now extinct. The remaining lions live in Africa, mainly around the Serengeti National Park in Tanzania and the Kruger National Park in South Africa.

Queen of the Jungle?

With its great mane of regal, golden fur, the male lion is an impressive sight. Yet it may be the smaller, un-maned females who are the true monarchs of the savannah. Naturally, lions are social animals and live in family groups called prides. Most of the pride, between 10 and 30 strong, are females, who are usually related through several generations. It's the lionesses who do all the hunting for the pride, although male lions do kill for themselves if the opportunity arises.

As a lioness can't run as fast as a cheetah or leopard, she has only a 30 per cent chance of a successful kill on her own, but by working in groups this real-life 'sisterhood' can double the odds. During the hunt, each lioness heads off in a different direction to surround the prey. Once in position, they crouch down in the long grass and stalk their victim until they're close enough to run it down. Using their powerful paws and body weight to pull the prey to the ground, they then suffocate it by biting its throat. Even working in groups, catching large prey is not an easy task, so lionesses usually hunt in the dark, when it is much easier to catch prey unawares. Yet, after the females have done all the work, the male will often scare off the other members of the pride to get himself 'the lion's share' of the feast.

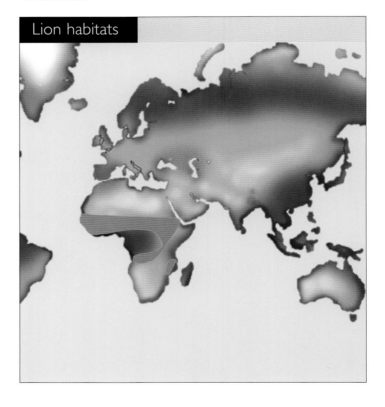

Lion habitats

Hunter or Scavenger?

A fully grown male lion can eat 40kg (88lb) of meat in one sitting. After he's gorged himself on a kill, he may not need to eat again for up to a week. In fact, for such powerful animals, lions are incredibly lazy. The size of an African Lion's territory varies depending on the size of the pride, from 20 to 400 square kilometres (7–154 square miles), but during an average day he will travel only about 8km (5 miles) and spend just a few hours hunting. A lion's preferred pastime is actually dozing in the sun – and he can spend up to 20 hours a day doing nothing else! Sometimes lions don't even bother to hunt for themselves. Hyenas have a reputation as scavengers who steal food from larger carnivores, but recent studies have shown that lions are just as likely to steal from hyenas, using their size to intimidate and chase the pack away from a fresh kill.

Truly Dangerous?

Lions are quite capable of killing humans, and when one went on the rampage in Malawi in 2003, it took four hunters to stop its fearsome killing spree. Even with a bellyful of bullets, and its intestines falling out, the lion still managed to maul two of the hunters before dying.

Despite their size and awe-inspiring roar, lions are not as dangerous as many of Africa's large herbivores. They are powerful and unpredictable, but they have learnt over the centuries to fear and avoid us. Lions may occasionally approach villages and settlements, usually to prey on livestock, but they usually stay well clear of humans. Sadly, the same is not always true of people – around 74 per cent of all lion deaths are still caused by humans and it's estimated that only about 23,000 of these spectacular animals still remain in the wild.

As evening approaches, a group of lionesses set out to hunt. Moving in a line, they use their keen sense to search for prey.

The tan colour of the lionesses allows them to blend in with the savannah grass. Using any cover available, they close in.

The prey are oblivious to the danger as the lionesses on the flanks of the line move forward, cutting off their escape route.

As the lionesses at the centre of the line charge, their prey are forced to flee directly towards the ambush.

Locust

Every continent has its own variant of this prolific insect, but the most dangerous to humans are the Red Locust, the Central African Desert Locust and the widely distributed Migratory Locust. Once on the move, swarms can travel so quickly that it takes only a few days for them to sweep all the way through Africa, across Europe and into southern Russia, bringing famine to entire continents as they migrate.

Key Facts	ORDER *Orthoptera* / FAMILY *Acrididae* / GENUS & SPECIES *Schistocerca*
Length	Up to 90mm (3¹/₂in); male smaller than female
Sexual maturity	After the fourth moult; the time this takes varies
Breeding season	Usually after the rain
Breeding Interval	Eggs can remain dormant in the ground for years
Number of eggs	5 or 6 pods of about 100 eggs each
Typical diet	Leaves, stems, flowers and fruits of wild plants and cultivated crops
Life Span	30–50 days as a hopper; 4–5 months as an adult

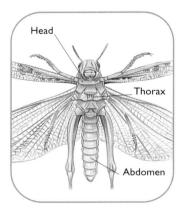

Body Structure
All adult members of the insect family have a similar body structure, comprising the head, thorax, and abdomen. This is all protected by a tough outer shell.

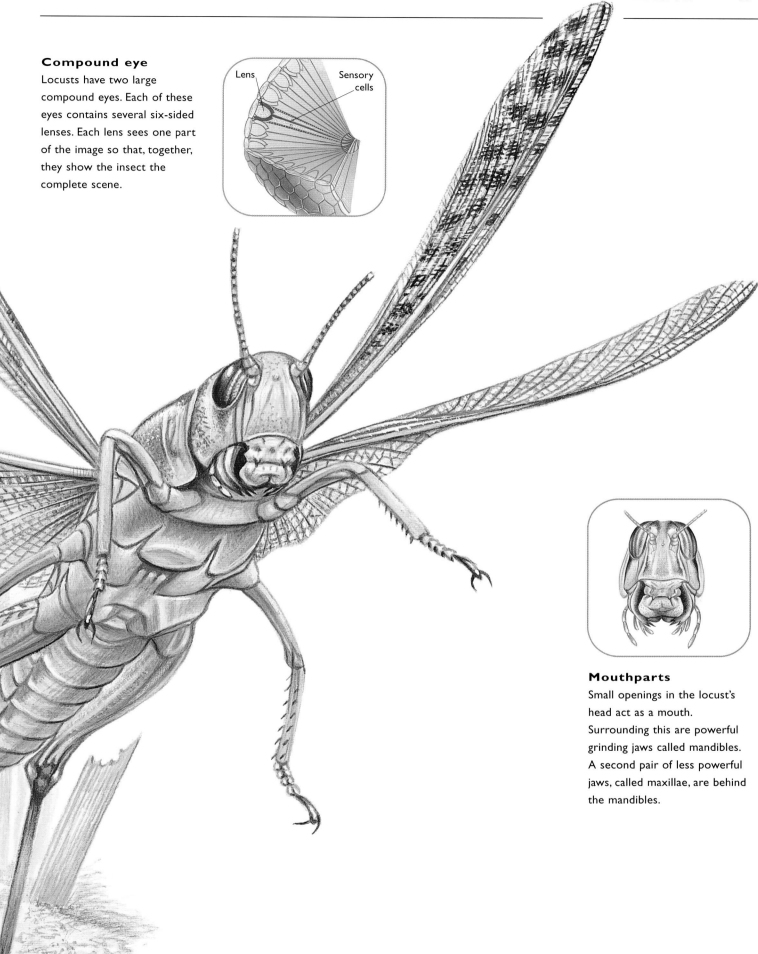

Compound eye

Locusts have two large compound eyes. Each of these eyes contains several six-sided lenses. Each lens sees one part of the image so that, together, they show the insect the complete scene.

Lens

Sensory cells

Mouthparts

Small openings in the locust's head act as a mouth. Surrounding this are powerful grinding jaws called mandibles. A second pair of less powerful jaws, called maxillae, are behind the mandibles.

Desert Locust habitats

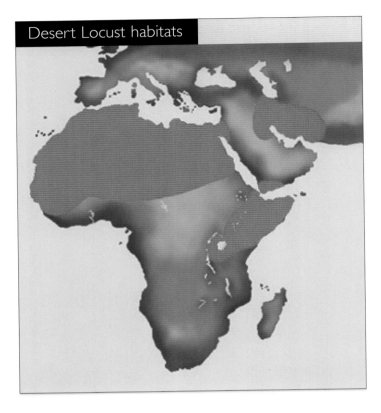

In the Old Testament, locusts were one of the ten plagues sent by God to punish the Egyptians, and even today many African farmers would recognize this description of the aftermath of a visit by this devastating insect: 'They covered the ground, until it was black; they ate everything …including all the fruit on the trees. Not a green thing was left…in all the land' (Exodus 10:15).

Adapt and Survive

Locusts are part of the insect family, which means that they have three pairs of legs, a body divided into three parts (head, thorax and abdomen) and a tough outer shell. Throughout the world, insects are the most successful group of animals: in every 2.6 square kilometres (1 square mile) of land, there are as many insects as there are people on the entire planet.

Part of the reason for this success is that they reproduce in huge numbers – incredibly quickly. Some insects can produce several generations of offspring within the year. This means that they can change quickly in response to environmental changes which might otherwise wipe them out, adapting themselves to live in almost any conditions, in any part of the world. Wherever you go, from the driest, waterless desert, to the coldest ice-bound tundra, you'll find a thriving insect population.

Insatiable Appetites

The name 'locust' refers to any type of migratory grasshopper. Locusts typically have short antennae (feelers) on their head; long hind legs for jumping; and four wings, which can be folded out of the way when not in use. Like all insects, locusts also have small openings in their head that act as a mouth. Surrounding this opening are their mouth-parts, which vary from insect to insect, depending on how it eats its food. Some insects, for example, suck up their food, using sharp needles called stylets to pierce leaves or animal bodies. Locusts, like termites, chew and cut their food. Using a set of powerful grinding jaws called

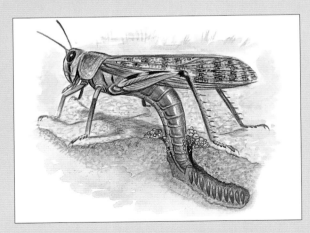

The eggs of the female locust are buried in a pod underground, then secured with a white foam that hardens into a protective seal.

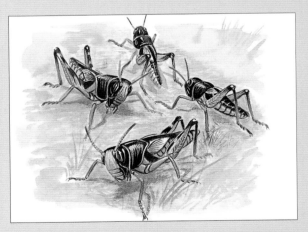

Once hatched, the maggots burrow to the surface, where they shed their skins, emerging as tiny hoppers.

mandibles, a locust can tear and rip its way through the toughest of materials. A second pair of less powerful jaws, called maxillae, are behind the mandibles. Both jaws work sideways, rather than up and down, making for an extremely effective eating machine!

Plague Proportions

No one knows for sure why locusts swarm, but it seems to happen when there's been plenty of rain and warm weather. Female locusts lay their eggs in soil. Damp, warm conditions encourage these eggs to hatch and also provide an abundant supply of green shoots for the young, emerging, locusts to eat. Once hatched, the wingless locusts, called hoppers, gather in small groups. In this phase, they're not a particular problem to farmers, but as numbers increase so do the dangers of the locusts swarming.

After about three weeks, the hoppers develop wings and the swarm takes to the air in search of food. Such locust swarms can be huge. One, in North Africa, was estimated

Comparisons

In Africa, the Desert Locust poses a constant threat to lives and livelihoods. It's especially dangerous because it often occurs in regions where food is already in scarce supply. A close second, in terms of destructive power, is the Migratory Locust. Although much smaller than its African counterpart, this sandy coloured insect covers a huge range, bringing destruction and devastation to farms in regions ranging from Africa to Northern Europe.

Migratory Locust

Desert Locust

to contain 150 billion insects, and covered an area measuring 121km by 26km (75 by 16 miles). 'You couldn't see soil any more', said one witness, 'all you could hear was chewing.' During a swarm, the locusts become so frenzied in their pursuit of food that they will attack anything that's green. They have been known to eat green clothes off clothes lines, attack green buildings or even make a meal of each other!

The immature hopper must shed its skin five more times before it becomes a fully developed, winged adult.

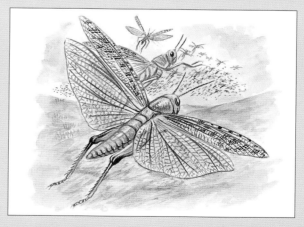

The locusts migrate in search of food once their rapidly increasing numbers have exhausted the local supply.

Nile Crocodile

The Nile Crocodile is the world's most terrifying freshwater predator. With voracious appetites, these armour-plated killers will attack anything that comes near the water's edge, using their huge jaws to crush bone and tear off hunks of flesh. As they are unable to chew, they use their bulk to drown large animals before storing their bodies in underwater larders to rot.

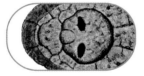 

Nostrils

With nostrils on the top of its snout, the Nile Crocodile can remain almost entirely submerged in water, and still breathe. Specially adapted flaps seal the nostrils when the crocodile dives.

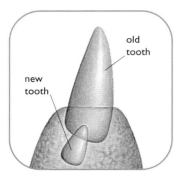

old
tooth

new
tooth

Teeth

A crocodile's teeth grow continually. Old, worn teeth are eventually replaced when new ones push them out from below.

Key Facts	ORDER *Crocodylia* / FAMILY *Crocodylidae* / GENUS & SPECIES *Crocodylus niloticus*
Weight	Up to 990kg (2180lb)
Length	Up to 6m (19ft 6in)
Sexual maturity	7–15 years
Breeding season	Varies with latitude, but coincides with dry season
Number of eggs	25–100
Hatching period	95–100 days
Birth interval	Probably annual
Typical diet	Mammals such as zebra, cattle and wildebeest, other reptiles (including other crocodiles), fish, birds, carrion; young Nile Crocodiles eat frogs and insects
Lifespan	70–100 years

Growing up to 6.2m (20ft) in length, these powerful green-grey reptiles have a long and well-deserved reputation as man-eaters. They were so feared in Ancient Egypt that people worshipped them in the form of the crocodile god, Sobek.

Ancient and Deadly

Crocodiles are reptiles, which means that they are cold-blooded. Unlike mammals, whose blood is heated inside the body by burning food for fuel, a reptile's blood is the same temperature as its surroundings. So to stay alive they need to keep warm and avoid extreme cold. Nile Crocodiles can, in fact, often be seen basking on the river banks to warm themselves in the early morning sun. In this un-warmed state, they're slow and sluggish. Nature, however, has given them other advantages to compensate. With eyes and nostrils on the top of their heads, they can sit almost completely submerged in the water, unnoticed, until prey approaches. Their throats have a slitlike valve, which closes when they're under water, so they can eat and breathe without drowning. They have fantastic eyesight, with vertical pupils that widen in the dark to aid night-time hunting. They also have a semi-transparent eyelid, called a nictitating membrane, which slides over the eye to protect it under water. Such adaptations make them particularly effective hunters.

Caring Parents

Nile Crocodiles may be fearsome killers, but they're dedicated parents. Like most reptiles, crocodiles lay eggs.

Nile Crocodile habitat

These look a little like hen's eggs but have a tough, leathery shell. Female Nile Crocodiles lay up to 90 eggs during the dry season. These are buried in sand or mud, and closely guarded until the rains come and they're ready to hatch. When this happens, the young crocodiles call out to their parents with a series of short grunts.

Unusually for the animal kingdom, both parent crocodiles take an active part in rearing the young so that,

Comparisons

American Alligators are closely related to Nile Crocodiles. They enjoy similar habitats, they're both skilled hunters, and they grow to a comparable size. Outside of their natural environment, in fact, they can be quite difficult to distinguish. The secret is to look at the business end – the mouth. When a crocodile closes its mouth, the fourth tooth in the lower jaw is visible. In an alligator, this tooth is hidden. The alligator's head is also shorter, rounder, and blunter.

Nile Crocodile

American Alligator

when it's time, either the male or the female may take on the job of digging up the buried eggs. Using their mouth, they help crack open the shells and carry the hatchlings to the water's edge. At this stage, the young crocodiles are only about 30cm (1ft) long. A fully grown Nile Crocodile has few natural enemies, but they are extremely vulnerable in these early months and often fall prey to monitor lizards, turtles and catfish. Within a few years, though, the tables will be turned and it will be the lizards and catfish who are on the menu.

Dangerous Neighbours

Nile Crocodiles have thick, waterproof, scaly skin, which is designed to prevent dehydration and loss of body salt. This skin has no osteoderms (bony plates), which means that it's ideal for clothing manufacture. In fact, crocodile skin has been used by humans for centuries to make tough, durable items such as boots. Today, much of the skin used by the fashion industry comes from crocodile farms. Many African countries have also signed international agreements to limit the number of crocodiles that they kill. This is good for the crocodiles, of course, but isn't always beneficial to the human population. Since Malawi signed the Convention on International Trade in Endangered Species (CITES), for example, there's been a dramatic increase in the number of humans killed by crocodiles – at least two a day in some areas. There are simply so many Nile Crocodiles in some rivers that the population is starving and some have been driven to eat the very people who are trying to protect them!

A silent approach is vital if the crocodile is to get close enough to attack its unwitting victim.

With incredible speed, the crocodile explodes from the water, clamping its powerful jaws around the leg of its startled prey.

Unable to escape, the prey is dragged into the waterhole as the crocodile prepares to drown it.

Although the crocodile has a strong bite, it is unable to chew. By rolling in the water, the crocodile not only drowns the zebra but is also able to rip meat from the carcass.

Puff Adder

The flat, brightly patterned head of the Puff Adder holds a lethal secret, for it's here that this deadly reptile's venom glands can be found. The venom may not be as toxic as the Black Mamba, but just four drops are enough to kill a grown man – and the Puff Adder can inject up to 15 drops in a single bite.

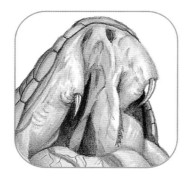

Fangs
A Puff Adder's fangs are set on 'elastic' hinges, which means that they spring forward when the snake opens its mouth, and fold back when it's closed.

Key Facts

ORDER *Squamata* / FAMILY *Viperidae* / GENUS & SPECIES *Bitis arietans*

Weight	1.5–2kg (3lb 5oz–4lb 6oz)
Length	70–150cm (28–59in)
Sexual maturity	About 2 years
Breeding season	Dependent on rainfall; usually during the early rains
Number of young	Usually 20–30; occasionally over 50
Incubation period	About 5 months; eggs hatch inside female
Birth interval	1 year
Typical diet	Rodents and other small mammals, birds, lizards and amphibians
Lifespan	Up to 15 years

Mouth
Poison stored in glands in
the upper jaw is fed to the
snake's hollow fangs by a
narrow tube.

Africa is home to eight species of Puff Adder, from Peringuey's Desert Adder, which is the smallest of the family, to the largest, the Gaboon Viper. This successful snake is so well adapted to surviving Africa's extremes that Puff Adders can be found throughout this vast continent.

An Invisible Killer

For a predator to be successful, good camouflage is essential. This is a lesson that members of the Puff Adder family have learnt well. On the dry sands and soils of Africa's deserts and savannahs, Puff Adders come in all shades of desert camouflage, from muted browns, to yellows and greys. These colours harmonize well with those of the environment, so that by lying still, or burrowing into the sand, a Desert Adder can become virtually invisible. A snake's colour comes chiefly from pigment cells in the layers of its skin. Most snakes have fairly muted colours, but an exception is the Rhinoceros Viper, so called from the upturned scales on its head, which look like horns. Indeed, this snake is positively showy, with scales of purple, blue, green, black and red, and it's difficult to imagine how such a brilliantly coloured animal could easily blend into the background. Yet, on the rainforest floor, among the fallen leaves and foliage, this skilled hunter can move around almost unnoticed.

A Deadly Development

All snakes are believed to have developed from a common lizard ancestor around 100 million years ago. During this process, they lost their legs, but gained a number of other important survival traits. Being carnivores, the most

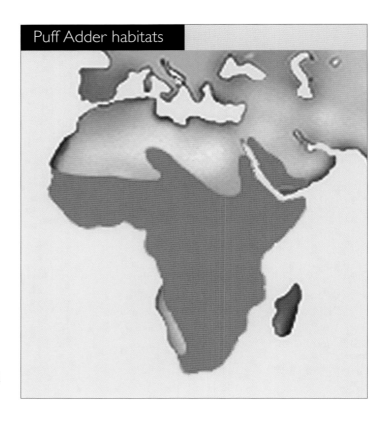

Puff Adder habitats

important of these developments was venom. As snakes don't have claws, they need a way to disable and kill their prey. Large constrictors, such as pythons and boas, suffocate their prey by throwing coils of their body around their victim and slowly crushing them. A quarter of all snakes, though, produce poison, using modified saliva glands. The delivery mechanism for this poison is its fangs, and, again, nature has come up with a few variations. Some snakes, for

Comparisons

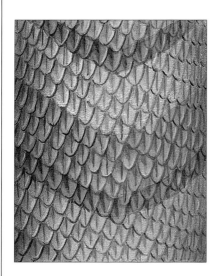

Puff Adder

The Gaboon Viper, like the Puff Adder, belongs to the viper family (*Viperidae*). In common with many members of this colourful and deadly group, Gaboon Vipers make their home in Africa, and are at their most comfortable amongst trees and foliage of the rainforest floor.

Vipers are typically short and thick in the body, with a wide head to accommodate their large venom glands and long fangs. They are also ovoviviparous, meaning that the female produces eggs which hatch within the body.

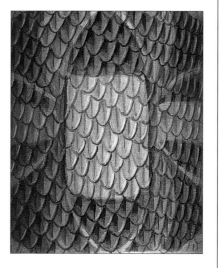

Gaboon Viper

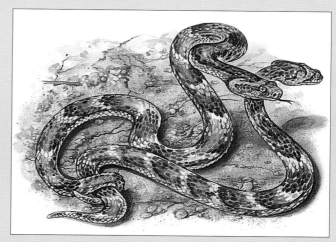

These two males are locked in a 'combat dance' over the right to mate with a receptive female nearby. This contest continues until one retreats.

For the victorious male to mate with the female, he must curl his tail underneath hers and insert his sex organs into a cavity at the end of her body, called the cloaca.

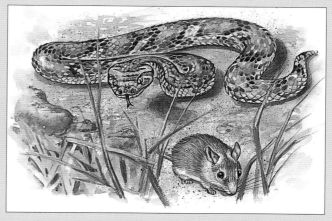

The gestation period lasts for about five months before the fully formed snakes hatch from their shells.

Soon after hatching, the young leave the parent to begin hunting for themselves.

example, have back-facing fangs at the rear of the jaw, which are used to inject poison and paralyse their prey as they swallow it. Others, like Puff Adders, have fangs at the front of the jaw. Moving the jaw forward allows the fangs to become erect so that the adder stabs the poison directly into its prey.

Swallowed Whole!

Puff Adders have an unusually loud hiss, which is made by forcing air through their lungs. This noise is a warning to stay well away. It takes a lot of resources to produce venom, and snakes would rather warn off large animals than waste poison on something they can't eat. Snakes can't tear or chew their food into smaller pieces, so a large

mammal is simply too big for a Puff Adder to handle. They do, however, eat a wide variety of smaller food, such as frogs, which can often be grabbed and eaten without the need to use poison.

With larger animals, like birds and lizards, the snake uses its fangs to inject the poison. Once the animal is dead, the adder swallows the animal whole. They can do this because their jaw is split from the skull at the chin, allowing it to flip right open on strong elastic ligaments. Backward-facing teeth help pull the corpse into the mouth, and then a series of muscle ripples force the corpse down the snake's throat. Once it has eaten, the full adder will often bask in the sun to raise its body temperature and speed up digestion.

ARCTIC
OCEAN

KARA SEA

Central
Siberian
Plateau

ASIA

Ural Mts.

SEA OF
OKHOTS

Lake
Baikal

Mongolia

Tien Shan Mts.

Gobi
Desert

Plateau
of Tibet

Himalayas

Asia

Asia is the world's largest continent, covering 44,936,000 square kilometres (17,350,000 sq. mi.). Within this vast social and cultural melting pot, 3.4 billion people – and thousands of species of animals – live out their lives in some of the world's most extreme environments.

~

In the east, sandwiched between China and Siberia, we find the vast Gobi Desert, home to numerous hardy animal survivalists. Travel west, and we come to the jagged pinnacles and deep valleys of the Pamir Knot, where Afghanistan, Pakistan and China collide. Of all the continents, Asia is the most mountainous. Indeed, this area of rugged natural beauty contains so many gigantic peaks that it's known as 'the roof of the world'. And here, isolated from the hustle and bustle of everyday life, many of Asia's rarest predators are able to make their homes.

Swing South and, gathered around the banks of the mighty Indus and Ganges Rivers, millions of Indians make their homes, living almost shoulder-to-shoulder with the great herbivores such as the elephant, fearsome hunter-killers like the tiger and venomous snakes such as the King Cobra, the world's largest. Finally, stepping off the South Eastern coast, we come to the Philippines. This group of 707 islands curves out across the South China Sea to form, with Borneo, Malaysia and Indonesia, a half-moon of tropical, volcanic islands that sit on the very edge of this colossal continent. On these islands are some of Asia's densest rainforests and richest, most fertile plains.

From fast and fearsome plains predators, to the camouflaged rainforest hunters, each region of Asia is home to its own spectacular species – all uniquely adapted to fight, thrive and survive.

Asiatic Black Bear

With one powerful blow from its mighty paws, an Asiatic Black Bear is able to bring down an animal the size of a small pony. Fiercer than the larger Brown Bear, this relatively small member of the family *Ursidae* (bear family) is nevertheless notoriously aggressive.

Foot

On each paw are 5 long, razor-sharp claws, which can't be retracted (pulled back). Instead, they are used to give the bear additional grip.

Teeth

As their teeth prove, Asiatic Black Bears are omnivorous – they eat animals and vegetables – so they have teeth that can both grind and tear.

Key Facts	ORDER *Carnivora* / FAMILY *Ursidae* / GENUS & SPECIES *Ursus thibetanus*
Weight	Male 50–120kg (110–265lb); female 42–70kg (92–154lb)
Length	Male 1.4–1.7m (4ft 7in–5ft 7in); female 1.1–1.4m (3ft 7in–4ft 7in)
Sexual maturity	3 years
Mating season	June to July
Gestation period	Up to 220 days (including delayed implantation)
Number of young	Usually 2
Birth interval	2–3 years
Typical diet	Succulent vegetation, fruit, nuts, insects and carrion
Life Span	Up to 25 years

Asiatic Black Bear habitats

With long fur around their shoulders and throat, their large furry ears and their lumbering gait, Asiatic Black Bears look rather like over-grown 'teddy bears'. These mighty mammals may seem cuddly, but appearances can be deceptive. They are actually skilled hunters and dangerous predators. Despite their bulk, angry Asiatic Black Bears are quick to respond to anything or anyone that threatens them or their young.

What's in a Name?

Worldwide, there are seven known main species of bear, which include the American and Asiatic Black Bear, the Brown Bear, Polar Bear, Sloth, Spectacled and Sun Bear. Giant Pandas are also sometimes classified as bears, although many zoologists believe that they are more closely related to racoons. Traditionally, all animals have a Latin name, which enables scientists to identify them clearly. Under this system, called 'taxonomy', the Asiatic Brown Bear is called *Ursus tibetanus*. Such complex names, however, tend not to be used in everyday life. The preference is for 'common names', which can be confusing because they vary from country to country. Asiatic Black Bears, for example, are known as Himalayan or Tibetan Bears, since they live in the mountains and hills of both regions. Sometimes they're also called Moon Bears, on account of the white, crescent-shaped mark on their belly, which helps distinguish them from American Black Bears, which are otherwise similar in size and appearance.

Big Appetites

In its natural environment of the forests and hills of southern and eastern Asia, an Asiatic Black Bear's preferred food is small mammals, reptiles and fish. Although more carnivorous than its American counterparts, it will supplement this diet during the year with eggs, berries, seeds, nuts and leaves. True to stereotype, they also love honey. Using its powerful paws, a hungry bear can easily tear a hive apart to get access to the honeycomb inside. Luckily, its thick fur prevents the bear from being seriously injured by the stings of angry bees.

Comparisons

The Spectacled Bear is the only bear in Latin America. Separated from its Asian cousin by the waters of the Pacific Ocean, this small bear probably arrived in the Americas millions of years ago, before the land masses divided. Yet despite the distance and timescale which separates these two species, they still share many traits. Though small, like all bears, the Spectacled Bear is agile and powerful. It's also an opportunistic feeder and enjoys a varied diet.

Spectacled Bear

Asiatic Black Bear

Most bears sleep during the winter in a cave or burrow. By eating lots of easily available, high protein, high calorie foods like seeds, nuts and honey, they are able to store enough fat in their bodies to prevent starvation during these barren months, when food may be scarce. It is believed that Asiatic Black Bears may not fully hibernate; even so, they sleep enough to need these additional reserves. This means that Asiatic bears in the wild enjoy a varied diet, eating food from both vegetable and animal sources. It's when black bears encounter unusual situations that they begin to abandon these 'normal' patterns of behaviour and conflicts with the humans flare up.

Under Threat

In China, Black Bears are killed for their gall bladders, which are regarded as an essential ingredient in many traditional medicines. In Taiwan, bear paws are eaten as a delicacy. In India, young bear cubs are often captured and forced to 'dance' for the public, due to their natural ability to balance on their hind feet. All over Asia, bears are under threat from a variety of sources, yet it's the humble farmer who is their greatest enemy. To a bear, a farm full of healthy young animals must seem like a convenience store. So it's no wonder that, when given the opportunity, Asiatic Black Bears will have no hesitation in killing and eating pigs, sheep and cows, despite the fact that these animals are much larger than their natural prey. Similarly, from a bear's point of view, an angry farmer protecting his stock is just a competitor trying to get between him and an easy meal. The result, throughout Asia, has been deaths on both sides, with the Black Bear increasingly the loser in this battle for land and resources.

Using its acute sense of smell, the bear follows the scent of honey to a nest of wild bees high in the trunk of a forest tree.

Although the bear is extremely heavy, its incredible strength and hooked claws mean it is well suited for climbing trees.

The bear is well protected against the stings of the bees by its thick coat as it plunders the hive. The honeycomb and bee larvae are rich in calories and protein.

The bear gorges on the sweet honeycomb it has stolen, ignoring the angry bees swarming around him.

Bengal Tiger

Tigers are the largest members of the cat family. Their sleek, muscular bodies and powerful limbs make them efficient hunters, but it's their strength and beauty that have made them one of Asia's most admired — and feared — great carnivores.

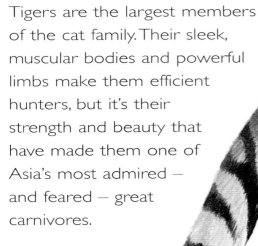

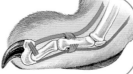

Teeth

Using its 4 pointed canine teeth like knives, the Bengal Tiger kills its prey efficiently.

Claws

Long claws are used to tear into and hold prey. When not in use, they can be pulled back into the paw to prevent them from becoming blunt.

Ears

White crescent-shaped patches of fur behind the tiger's ears help it to easily identify other tigers when they are hunting under the cover of darkness.

Key Facts	ORDER *Carnivora* / FAMILY *Felidae* / GENUS & SPECIES *Panthera tigris tigris*
Weight	180–265kg (397–584lb)
Length Head & Body Tail	1.9–2.2m (6ft 3in–7ft 3in) 80–90cm (31–35in)
Shoulder height	90–95cm (35–37in)
Sexual maturity	Female 3–4 years; male 4–5 years
Mating season	Winter to spring
Gestation period	95–112 days
Number of young	2 to 4
Typical diet	Sambar deer, chital deer, buffalo, wild pigs, gaur and monkeys
Life Span	Up to 26 years in the wild

Comparisons

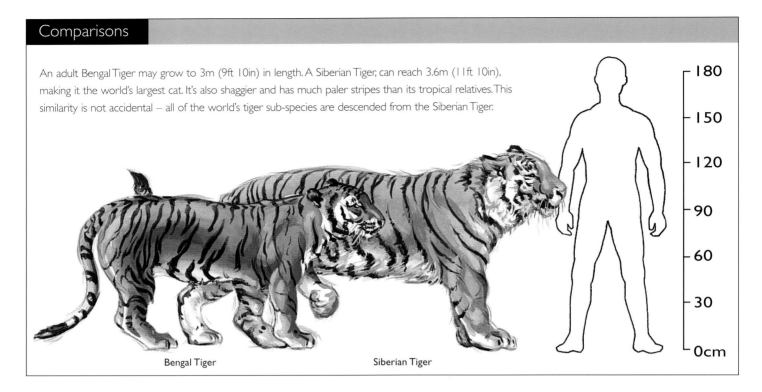

An adult Bengal Tiger may grow to 3m (9ft 10in) in length. A Siberian Tiger, can reach 3.6m (11ft 10in), making it the world's largest cat. It's also shaggier and has much paler stripes than its tropical relatives. This similarity is not accidental – all of the world's tiger sub-species are descended from the Siberian Tiger.

Bengal Tiger   Siberian Tiger

It was not until the eighteenth century, when the British began to settle in India, that Westerners encountered the tiger. Their horror of, and fascination with, this renowned man-killer was summed up by the poet William Blake (1757–1827) in perhaps his best-known work, 'The Tyger'. In it, he speculated whether God could really have created such a beautiful yet fearsome creature:

> Tyger! Tyger! burning bright
> In the forests of the night;
> What immortal hand or eye
> Could frame thy fearful symmetry?

For centuries, such awe and fear has fuelled the arguments of both hunters and conservationists, who now hold the future of this magnificent animal in their hands.

Dangerous Liaisons

Male Bengal Tigers live and hunt alone. When they meet other tigers, they'll often rub their heads together in a friendly greeting and will even, occasionally, share a kill, but, like cheetahs, they're generally solitary. It can take up to two years before a tiger cub is able to feed itself. Until then, it relies on its mother, who occasionally hunts with her mate. Having a male around, though, can be dangerous. In common with many members of the animal kingdom, male tigers will sometimes kill a female's cubs if they are the offspring of another male. This is to ensure that his genes, not a competitor's, are carried on by the

next generation. Tigresses can give birth to as many as six cubs at a time, but these early years are so hazardous that usually only a few make it to adulthood.

Once fully grown, tigers are opportunistic feeders, eating anything from small mammals to farm animals, if they are easily available. In general, tigers prefer large prey – they can eat up to 30kg (65lb) of meat at one sitting and can easily bring down a full-grown ox.

Bengal Tiger habitats

Hunting and Fishing

Tigers live on the edges of forests and swamp lands and leave the safety of the trees only at night when they hunt. Using its sharp eyesight and keen sense of smell to track an animal's movements, a Bengal Tiger must time its attack for maximum effect, as it is capable of only short bursts of speed before it's too exhausted to continue the chase. Bounding out of cover, a tiger will grab its victim on the rump and use its claws and weight to pull it to the ground. Once the prey is dead, the tiger will often move the corpse to a hiding place, where it can return later to continue feeding.

With powerful shoulders and forearms, a healthy Bengal Tiger is capable of dragging a 230kg (507lb) deadweight over 500m (550 yards). If large prey is scarce, however, the tiger is just as happy hunting at the water's edge as in grassland. Like all cats, tigers are extremely nimble and, using their paws, make agile fishermen.

Fear and Loathing

Of the world's eight types of tigers, three probably became extinct recently, during the twentieth century. Today, the Bengal Tiger is on the endangered list. People have been killing tigers for centuries. In William Blake's time, thousands were killed by hunters, who kept their skins and heads as trophies. More recently, tigers have been trapped for their fur or body parts, which are used in traditional Chinese medicines.

Yet it's fear that remains the prime factor in the continuing fall in tiger numbers. Tigers do kill humans, but usually only when they're too old to hunt quicker animals or their natural prey is scarce. In Bangladesh, which is one of the world's most densely populated countries, the number of people killed by Bengal Tigers has risen rapidly over the last few years. As people start to clear forests and mangrove swamps to build farms, the tiger has no choice but to defend itself – and its territory – or die.

The buffalo calf is unaware of the danger because the tiger keeps downwind and uses the long grass as cover.

Closing to within striking distance, the tiger pounces. Finally seeing the danger, the startled calf attempts to flee.

Its powerful hindlegs driving it forward, the tiger strikes at the calf's neck with its razor sharp foreclaws.

If the calf's spine is not snapped by the initial blow, the tiger kills it with a bite to the throat.

King Cobra

Throughout the world, snakes are represented in mythology as the very embodiment of evil. Ancient people feared and respected them, and much of our instinctive dislike of snakes is based on this ancient dread — with good cause. They kill half a million people every year, and the world's largest venomous snake, the King Cobra, carries enough poison to kill 20 people in just one bite.

Key Facts	ORDER *Squamata* / FAMILY *Elapidae* / GENUS & SPECIES *Ophiophagus hannah*
Weight	5.5–8kg (12lb 2oz–17lb 10oz)
Length	Up to 4.9m (16ft); average 4m (13ft 1in)
Sexual maturity	About 4 years
Mating season	January
Number of eggs	18 to 51, usually 40 to 50; laid about 2 months after mating
Incubation period	70–77 days
Birth interval	1 year
Typical diet,	Mainly snakes, plus some lizards
Life Span	20 years on average

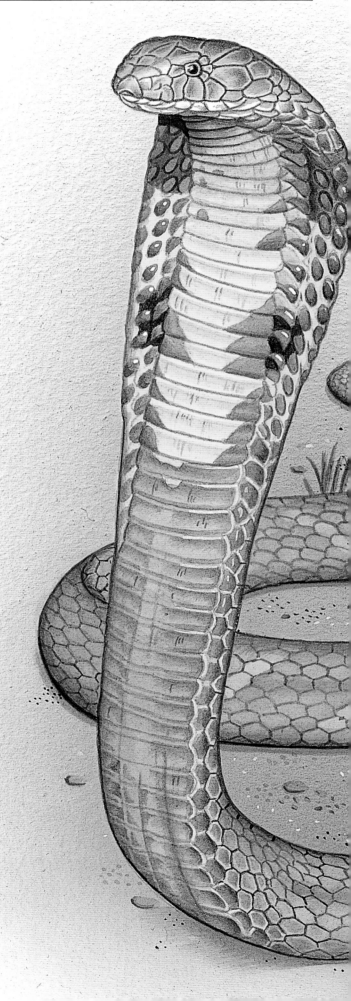

Teeth
Cobras fangs have a groove along the front edge, which is used to inject poison into their victims. In some species, this is a hollow tube.

Tongue
When a snake flicks its tongue in and out, it's actually collecting scent molecules from the air. These help it to detect its prey.

Hood
By moving its short neck 'ribs', the cobra creates a 'hood' around its head, which is designed to make it look bigger to potential enemies.

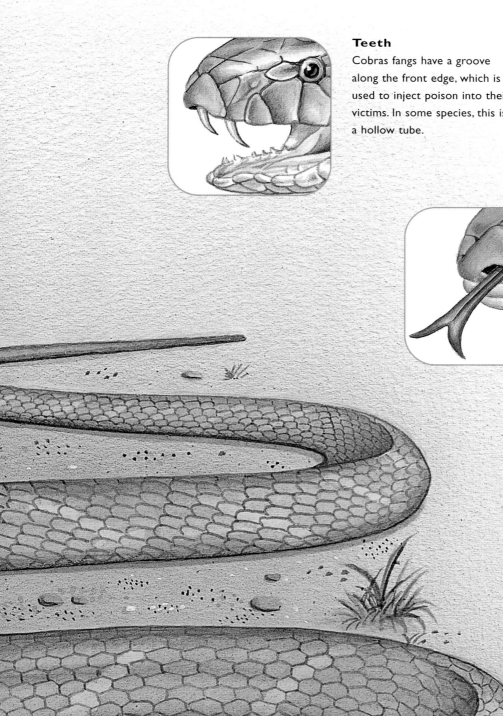

Comparisons

Death Adder

Green Mamba

King Cobra

Coral Snake

Cobras, kraits and mambas belong to the Family *Elapidae*. Members of this family have short, fixed fangs, unlike vipers, who have long fangs that spring forward when the snake opens its mouth. Of the 200 or so

Elapids, all carry poison, but not all boast about it. Unlike the Coral Snake which carries bold warning colours, Death Adders and Green Mambas are designed to blend into their surroundings.

Swift and deadly, the King Cobra is one of Asia's most unwelcome guests. Lethal to humans and animals alike, cobras rank as one of the continent's most dangerous and widespread snakes.

A Cruel Trick

If you've ever seen an Indian 'snake charmer' at work, you might think that these street performers are blessed with nerves of steel. In fact, the 'trick' they perform may look dangerous, but the charmers actually use the cobra's natural defences as part of their act. Snake charmers carry cobras around in large, straw baskets. At the start of the performance, the charmer takes the lid off the basket and the snake is momentarily overwhelmed by the light. In response, it rears up its body in a defensive position and, by moving its short neck 'ribs', creates a 'hood' around its head, which is designed to make it look bigger and more threatening to potential enemies.

A snake's eyesight is poor and it has no ears, so when the snake charmer begins to play his pipe, the cobra can't actually hear or see much of what's happening. The reason the cobra look likes it's dancing is because the snake is following the pipe's movement. This is bold enough for the snake to be able to focus on. By swaying backwards and forwards in time with the pipe, it keeps the potential threat within striking distance. The snake charmer is in no real danger, however, as most of the cobras used in this act have had their fangs ripped out and their mouths sewn up.

Advantages and Disadvantages

It would seem that being almost deaf and blind, cobras would make poor hunters. Yet, as is so often the case, nature has compensated for these apparent deficiencies. King Cobras are, in fact, able to 'sense' movement and sounds made by their prey by touching their jaws to the ground. Combined with visual information, this enables them to judge the distance between themselves and their

King Cobra habitats

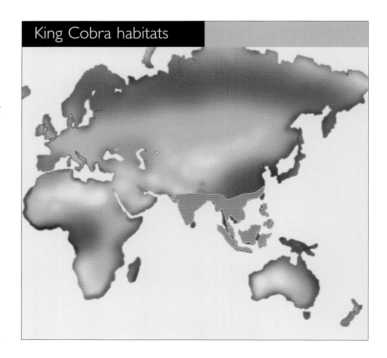

prey fairly accurately. Its other great advantage, of course, is its poison. When a King Cobra bites, it tends to hold onto its prey as long as possible. Compared to the Puff Adder, a King Cobra's fangs are quite small, so it actually chews down on its prey to force poison into the wound. The amount of damage inflicted depends on how long it's able to hold onto its victim. Some snakes' venom affects the victim's skin and muscle tissue, but a cobra's venom acts directly on the nervous system to cause paralysis. So, even if the snake is unable to kill its prey outright, it's likely to disable it enough to finish the job later.

Swift and Deadly

A King Cobra's favourite food is other snakes, but it will attack and kill much larger mammals if the opportunity arises. When roused, they have even been known to attack elephants, although, presumably with no intention of eating them! Female cobras, especially, can be irritable and volatile. They're protective mothers and it's during the breeding season that they're at their most dangerous. Generally, even the largest King Cobra, which can be as much as 2.4m (8ft) long, would rather back away from an unwanted conflict, but females will often strike out without provocation when defending their eggs. Throughout India, a third of all deaths from snake bites are caused by cobras, often because someone has strayed too close to a nest site. Careless or unwary humans are likely to get a very nasty surprise; without prompt treatment, a bite from a King Cobra can cause an agonizing death in just 13 minutes.

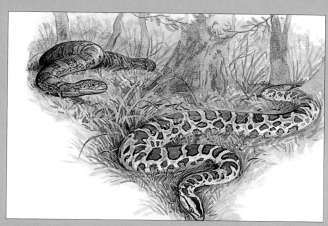

Using its tongue to gather scent samples from the air, the King Cobra discovers a python in the vicinity. When close enough, it tracks its prey by sight.

The python does not detect the cobra stealthily approaching through the grass. When it's within striking distance, the cobra coils its forequarters, ready to strike.

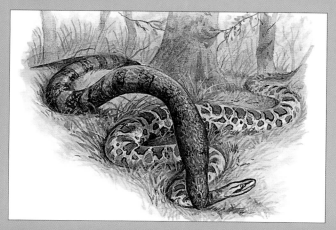

The cobra bites the neck of the python and injects its victim with venom in a single, lightening fast strike.

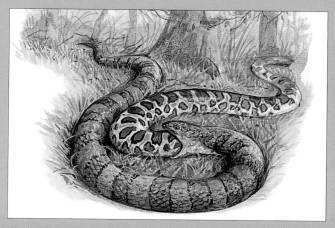

Paralysed by the venom, the python is powerless to resist as the cobra dislocates its jaws and carefully consumes it whole.

Komodo Dragon

The Komodo Dragon is a living reminder of the days when the great 'thunder lizards' – the dinosaurs – ruled the earth. A primitive and voracious reptile, the Komodo Dragon will attack and devour its own kind as well as the carcasses of any other animal.

Key Facts	ORDER *Squamata* / FAMILY *Varanidae* / GENUS & SPECIES *Varanus komodoensis*
Weight	Up to 165kg (363lb 12 oz)
Length	Up to 3m (9ft 10in)
Sexual maturity	5–6 years
Breeding season	All year; usually June to July
Number of eggs	4 to 30
Incubation period	32–36 weeks
Breeding interval	Annual
Typical diet	Carrion, scraps; a wide variety of prey, from insects, lizards, snakes and birds to rats, wild pigs, deer and water buffalo
Lifespan	About 50 years

Tongue
A striking feature of the Komodo Dragon is its incredibly long, slim forked tongue, which can be extended well beyond the mouth.

Teeth
With grating, serrated edges, a Komodo Dragon's teeth are easily able to slice through skin and muscle and crush bones.

Comparisons

Africa's answer to the mighty Asian Komodo Dragon is yet another member of the monitor family (*Varanidae*). The powerful Nile Monitor is dark in colour, with bands of spots around its body. Like all monitors, it is stocky, with stout, squat legs, and a small blunt head.

Ideally adapted for life along the river's edge, the Nile Monitor's tail is crested, to give it greater manoeuvrability in the water. The tail is also used as the weapon of choice against any would-be aggressor.

Komodo Dragon

Nile Monitor

Despite being the world's largest species of lizard, the Komodo Dragon was not discovered by the scientific world until 1912. Surviving dragons are now found on only a few small islands in the Indonesian archipelago: Rinca, Gili Motang, Nusa Kode and Komodo Island, which gave this brutal-looking lizard its name.

Feeding Time

Up until 1996, one of Komodo Island's most popular tourist attractions was a particularly gruesome demonstration of the awesome power of their most famous inhabitants. Several times a day, a goat would be left as a 'gift' for the dragons.

Lizards lack the in-built temperature control that warm-blooded animals have. This means that they are slow and sluggish in cold weather, but in the tropics, they thrive on the warm temperatures. Once they've been literally warmed up by the sun, Komodo Dragons can move at incredible speeds. Lifting their huge bodies off the ground in much the same way as a crocodile, the dragons propel themselves forward using their powerful forearms. Once a pack of komodos had caught its scent, the goat didn't have to wait long for visitors!

Voracious Appetites

When humans first saw the great Komodo Dragon, they must have felt as if they'd stepped back in time. Head to

tail, this solid, blunt-nosed lizard can grow to 2m (6ft 7in) in length and weigh a massive 250kg (550lb). To fuel such a huge body requires a regular input of food. Luckily, Komodo Dragons are not fussy eaters. They have an excellent sense of smell and the stench of rotting carrion will very quickly attract a pack of hungry dragons.

Using their sawlike teeth to tear the carcass apart, komodos can eat up to 80 per cent of their body weight in one sitting. With live prey, their approach relies more on

Komodo Dragon habitats

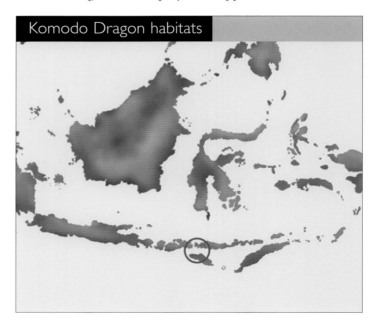

physical prowess than on stealth or cunning. They're not great hunters, but they don't need to be. Once within reach of their victim, they use their claws and body weight to pull their prey to the ground and overpower it. A fully grown Komodo Dragon can easily eat a goat or a pig, and has no hesitation in taking on much larger prey such as horses or buffalo. Like their reptile cousins, the snakes, they can 'disarticulate' (separate) their jaws, enabling them to swallow huge hunks of meat whole. So size is no guarantee of safety when a dragon's around. It just means that there's more for them to get their teeth into!

Primitive Poison

If you're lucky enough to get close to a Komodo Dragon, and live to tell the tale, one of the first things you'll probably notice is its terrible bad breath. Komodo Dragons

have a mouthful of bacteria, which they use as a primitive form of venom. So far, about 62 different types of bacteria have been identified. Most are so virulent that, to kill, a dragon has only to bite its victim a few times and wait for the bacteria to do their deadly work. It may take several days for infection to set into the wounds and kill an especially large animal, but a komodo bite can down a buffalo in less than eight hours.

If humans are bitten by komodos – and escape serious injury – there's still a real threat of death by septicaemia (blood poisoning) if the victim isn't treated quickly with antibiotics. The symptoms of blood poisoning include severe fever and, in extreme cases, the patient's blood will lose its ability to clot, causing excessive bleeding and, ultimately, death. These dragons may not belch fire, but their breath can still be just as deadly.

Smelling prey close by, the Komodo Dragon lies in wait, hiding in the long grass before bursting forth and attacking the pig.

Caught unawares, the pig is helpless as the komodo delivers a powerful bite to the neck.

Fatally wounded, the pig tries to escape. The Komodo Dragon lets it flee, as it will soon succumb to shock and blood loss.

The komodo tracks the dying pig and devours the carcass. Other Komodo Dragons are attracted by the scent of blood.

Leopard

Powerful and graceful, leopards are perhaps the most athletic members of the cat family, being equally skilled at running, jumping, climbing and swimming. Add to this their spectacular natural camouflage, and it would seem that the leopard is a natural-born hunter.

Jaws
With extremely powerful jaws, a fully grown leopard can bite down on its prey with tremendous force. This pushes its long canine teeth through flesh with ease.

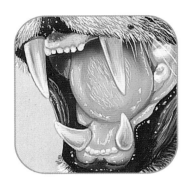

Paws

Leopards have relatively small paws compared to lions. Yet their claws are extremely long and make a very effective addition to their hunting kit.

Eyes

Like all cats, leopards have superb eyesight. Leopards are night-time hunters and their night vision is almost 7 times better than a human's.

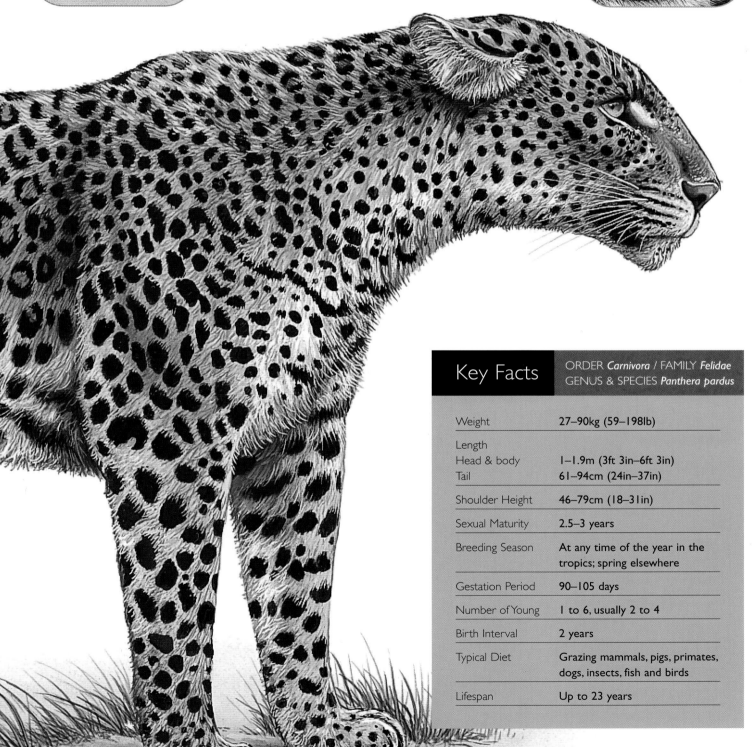

Key Facts	ORDER *Carnivora* / FAMILY *Felidae* GENUS & SPECIES *Panthera pardus*	
Weight	27–90kg (59–198lb)	
Length Head & body Tail	1–1.9m (3ft 3in–6ft 3in) 61–94cm (24in–37in)	
Shoulder Height	46–79cm (18–31in)	
Sexual Maturity	2.5–3 years	
Breeding Season	At any time of the year in the tropics; spring elsewhere	
Gestation Period	90–105 days	
Number of Young	1 to 6, usually 2 to 4	
Birth Interval	2 years	
Typical Diet	Grazing mammals, pigs, primates, dogs, insects, fish and birds	
Lifespan	Up to 23 years	

Comparisons

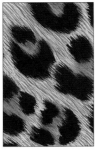

Leopard

Melanistic Leopard

Snow Leopard

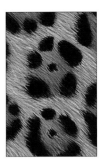

Jaguar

Leopards come in coats of many colours. As might be expected, the fur of a Snow Leopard is pale grey with splashes of black. This provides ideal camouflage in the icy wastes of their traditional home in the plateaus of Tibet. Asian Leopards have a golden, sandy coloured coat, with similar black patches. This might not seem like an ideal camouflage for grasslands. However, during the Summer, green grasses become scorched and dry, which means that the leopard's fur is an ideal match.

In the leopard, power and beauty have been combined to make this skilled carnivore into one of Asia's most striking-looking animals, a fact that sadly has not been missed by generations of hunters, who have killed this beautiful creature for its fur. Ranging throughout India and across to China, most sub-species of leopard, including the spectacular Snow Leopard and Clouded Leopard, are now endangered.

Far and Wide

Leopards are among the most widespread of the family *Felidae*, with sub-species from the mountains of China to the plains of Africa. When groups of animals become isolated from the main gene pool, small differences in the group may persist and can become more pronounced. This means that leopards vary not only from continent to continent, but from region to region. In general, the most noticeable differences are in their spots. On the leopard's legs and head, spots tend to be scattered randomly, whereas on the body they are gathered in little rosettes. These spots tend to be smaller on an African Leopard than on an Asian Leopard, while those on an Indian Mountain Leopard are further apart than on one that comes from the plains. Like human finger-prints, no two leopard coats are ever exactly the same, and in the past attempts were made to classify different varieties of leopard according to their spot patterns! It's a popular saying that 'a leopard can never change its spots', but some leopards don't even have spots. Black Panthers, for example, are actually melanistic (dark) leopards.

Learning to Kill

Adult females come into oestrus – that is, become sexually receptive – every 60 days for between seven

Leopard habitats

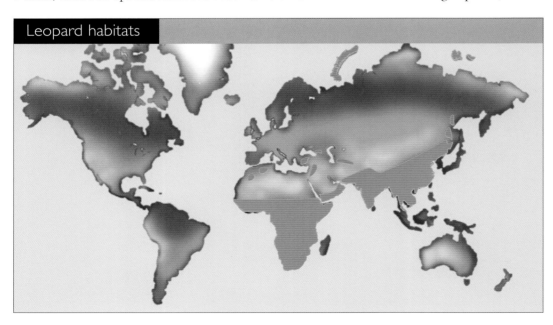

and ten days at a time. Once in oestrus, they will attract a mate using scent and calls. During this period, males may fight to win the right to mate with the female and injuries are common during such contests. Once she is pregnant, though, the male leaves and the female takes on full responsibility for raising and teaching her young charges. Typically, two or three cubs will be born, but anything up to eight are possible.

These tiny cubs will stay with their mother for up to 18 months, until they are ready to find a territory of their own. During this time, play is an essential part of the learning process. Cubs won't be allowed to hunt until they're about nine months old, but they will engage in play fights and mock hunts long before that. A leopard's life can be a short and a violent one, so learning the skills of survival at an early age is vital.

Super Cat

Leopards prey on about 92 different species, from dung beetles to young giraffes and, occasionally, humans. Their success is a result of numerous natural advantages. Their claws, for example, are extremely long (25mm/1in) and sharp. Their teeth are stronger, relative to their size, than a lion's, with four dagger-like canines that are used to puncture flesh and deliver the killing bite. Leopards are night-time hunters and their eyesight is almost seven times better than a human's. Yet it is their remarkable strength and agility that makes leopards so dangerous. A leopard can jump a distance of 4.5m (15ft) to bring down its prey. Once the killing bite is delivered, the leopard will hide the carcass up a tree, where it can eat undisturbed. Amazingly, it is estimated that a fully grown leopard can haul three times its own body weight up a 6m (19ft 7in) tree.

The leopard rarely attacks over open ground, preferring to use cover to get close to its prey.

Whether pouncing or making a surprise dash, the leopard strikes with a bite to the back, killing the impala.

Because the leopard will not consume the prey all at once, it drags it into a tree, out of the reach of scavengers.

With the carcass safely stored in the bough of the tree, the leopard can return and feed at leisure.

Reticulated Python

Unlike African Pythons, which prefer open ground, Reticulated Pythons are equally at home on the ground, in the treetops and in water. Silent and deadly hunters, these massive snakes are happy to make a meal of almost any animal.

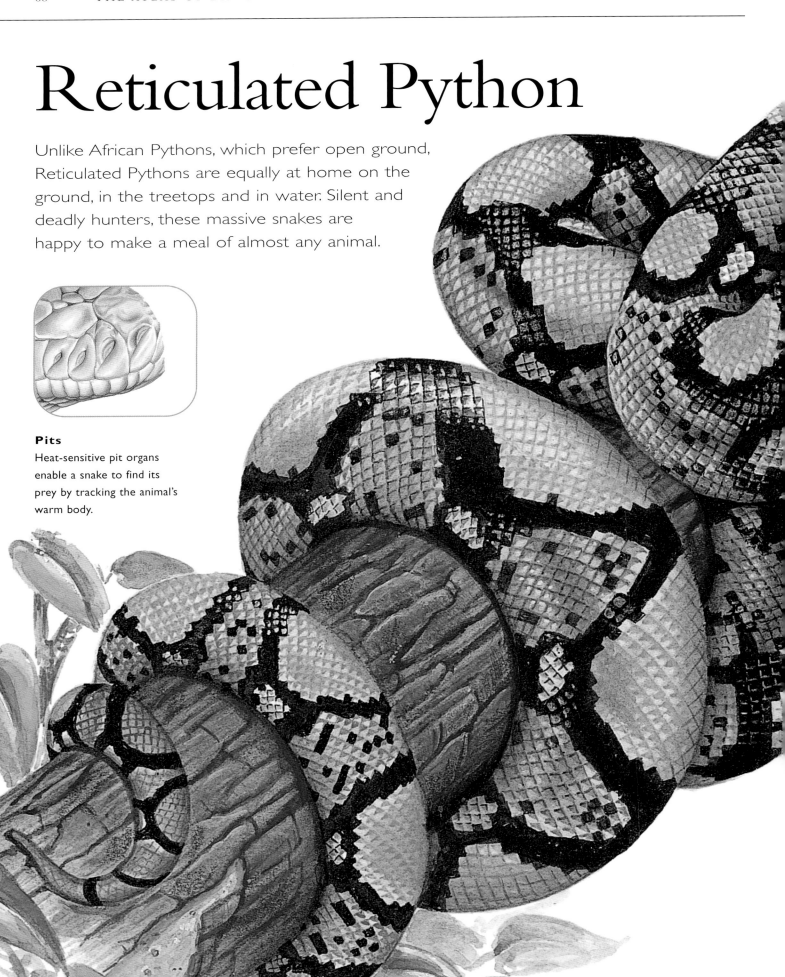

Pits
Heat-sensitive pit organs enable a snake to find its prey by tracking the animal's warm body.

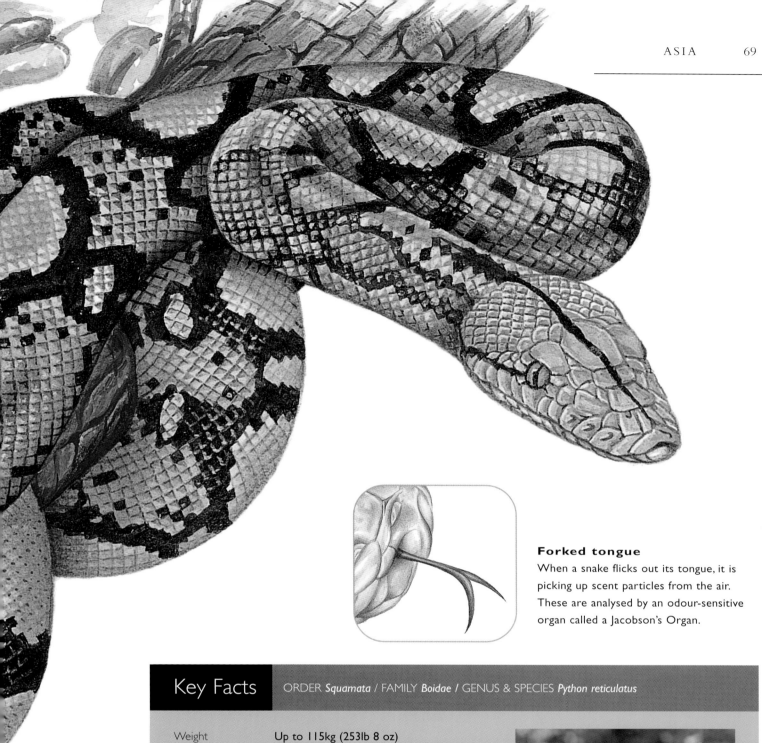

Forked tongue

When a snake flicks out its tongue, it is picking up scent particles from the air. These are analysed by an odour-sensitive organ called a Jacobson's Organ.

Key Facts

ORDER *Squamata* / FAMILY *Boidae* / GENUS & SPECIES *Python reticulatus*

Weight	Up to 115kg (253lb 8 oz)
Length	Average adult 5–6m (16ft 5in–19ft 7in) Occasionally reaches 9m (29ft 6in)
Sexual maturity	5 years
Breeding season	Variable
Number of eggs	Up to 100, but usually 20 to 50
Gestation period	10–12 weeks
Hatching period	11–13 weeks
Breeding interval	1 year
Typical diet	Small mammals, birds and reptiles
Lifespan	Unknown in wild; up to 27 years in captivity

Herpetologists estimate that there may be as many as 2700 species of snakes. Of these, the pythons are most obviously descended from 'legless lizards', since, at the tip of their stout bodies, they have small stumps, which are in fact 'vestigial' hind legs that have reduced in size and usefulness over many millennia.

Bouncing Babies

Reticulated Pythons are amongst the largest members of the family *Boidae*, which includes South American Anacondas and Boa Constrictors. No-one knows for certain how long snakes can live in the wild; there are accounts of some reaching 70 years of age, but it is thought that 20 is more usual. Snakes continue to grow their whole lives, so it's not unknown for these large reptiles to reach up to 9m (30ft) in length. Remarkably, much of this growth occurs during the first four years of the snake's life.

Other members of the family *Boidae*, such as Boa Constrictors, produce live young. The female Reticulated Python, however, is unusual in that she not only lays eggs but broods (sits on) them for three to four months. Using her body as an incubator, she constricts her muscles to produce heat that keeps the eggs warm. This increases the chances of a successful hatching, and baby Reticulated Pythons are a healthy 60–80cm (2ft–2ft 6in) long at birth. They can treble in length in their first few years of life!

Here Be Dragons!

The word 'dragon' comes from the Ancient Greek for snake, and it's likely that the fantastical creature of legend is based, not on a large lizard, but a python. In *The Historie*

Concealed in a safe spot on the forest floor, the female Reticulated Python lays a clutch of between 20 and 50 eggs.

To protect them from predators, the female remains wrapped around her newly laid eggs until they hatch.

The baby snakes have evolved a sharp 'egg-tooth' on the snout, specifically to help break the egg shell when their 90-day gestation period is at an end.

The newborn snakes must then fend for themselves with no further help from the mother. Although they are fully developed, they measure only 60-75cm (24–30in) in length.

of Serpentes, written in 1608, Edward Topsell describes how a dragon kills an elephant:

'They hide themselves in trees…In those trees they watch until the Elephant comes to eat and croppe off the branches, then suddainly, before he is aware, they leape… and…with their tayles or hinder partes…vexe the Elephant until they have made him breathlesse, for they strangle him with they'r foreparts.'

Pythons are constrictors, which means that they squeeze their prey to death. While it's unlikely that one could actually kill an elephant, the description of the dragon's technique is very reminiscent of the python's famous suffocating grip.

A Big Hug

Pythons are natural jungle dwellers. In fact, their prehensile tail allows them to climb trees with ease, while the netlike (reticulated) patterning on their skin makes them almost invisible amongst the leaves. Yet they seem to have a special liking for human settlements. In the past, they were regular night-time visitors to the streets of Bangkok – where numerous cats, dogs and rats offered the opportunity of an easy meal.

When hunting in its natural environment, a python may lie motionless for many hours until its victim is within reach. Using its jaws to hold its prey, the python will then wrap coils of its body around the captive animal, and slowly begin to tighten its hold. As its prey breathes in, the

Reticulated Python habitats

python pulls its coils ever tighter until its victim eventually suffocates. Like all snakes, the Reticulated Python can swallow animals much larger than itself: up to an astounding 54kg (120lb). There have even been a few reliable accounts of pythons eating leopards, an 80kg (176lb) bear and, in a few instances, human children. Once fed, the python's stomach and skin stretch to accommodate the meal. It may take several days for a large feast to digest before the snake disgorges (throws up) a ball of fur, which is the only part of its victim it cannot eat.

Comparisons

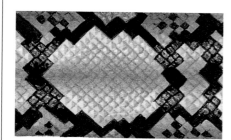

Reticulated Python

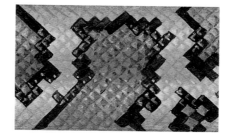

Indian Python

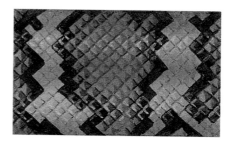

Burmese Python

Snakes get their colour from pigment cells, which lie deep beneath the layers of the snake's skin. Some coloration is also due to the way that the scales, which cover their bodies, absorb or reflect the light. Reticulated Pythons are so called because of their distinctive skin patternation, which is supposed to resemble a net. One of the simplest ways to identify different species of snake is by looking at their coloration. As these images show, there are obvious differences even amongst members of the same family.

Mongoose

Using its natural agility and guile, the mongoose preys on one of Asia's biggest killers – the snake. Amongst the trees, hills and grasslands of Asia, this small and daring animal has become the hunter of hunters.

Hindfoot
Dextrous, clawed toes give the mongoose additional grip. These are covered with thick, bristly hair, except on the very tips of the digits.

Key Facts	ORDER *Carnivora* / FAMILY *Viverridae* / GENUS & SPECIES *Herpestes spp*
Weight	1.7–4kg (3lb 11oz–8lb 13oz)
Length Head & Body Tail	48–60cm (19–24in) 33–54cm (13–21in))
Sexual Maturity	1–2 years
Mating Season	Varies according to region
Gestation Period	42–84 days
Number of Young	1 to 4
Typical Diet	Small mammals, birds, reptiles, invertebrates and frogs
Lifespan	12 years (in captivity)

Teeth
Sharp, piercing canines at the top and bottom of the jaw seize and hold prey. Carnassial teeth (just behind) cut and tear into flesh.

When the Indian-born author Rudyard Kipling (1865–1936) told his story of the great war that the mongoose Rikki-tikki-tavi 'fought single-handed through the bathrooms of the bungalow in Segawlee cantonment' against the evil cobras Nag and Nagainaan, he brought this tenacious little animal to the attention of the western world for the first time. Throughout Africa and Asia, however, people have long appreciated the skills of this 'giant killer' of the animal kingdom.

Pest Control

Mongooses have been trained and kept as pets by humans for thousands of years. Before the introduction of cats to Ancient Egypt, Egyptian Mongooses were kept in homes to protect the occupants from the threat of snakes. They were so revered that their tiny mummified bodies have even been found in temples and tombs. For centuries, they were used in the same way in India. In the nineteenth century, settlers took them to Jamaica, Cuba and Hawaii as a form of pest control. Unfortunately, mongooses don't just eat snakes, but thrive on a diet that includes a variety of small reptiles, birds and eggs, which they break by throwing at rocks. Once the industrious mongooses had worked their way through the viper population of Jamaica, they began to pose a serious threat to other indigenous species. Mongooses are such efficient predators that many countries now forbid their importation, since they can devastate local wildlife once loose.

Different, yet the Same

Mongooses belong to the family *Viverridae*, which includes civets and genets. Typically, members of this group share

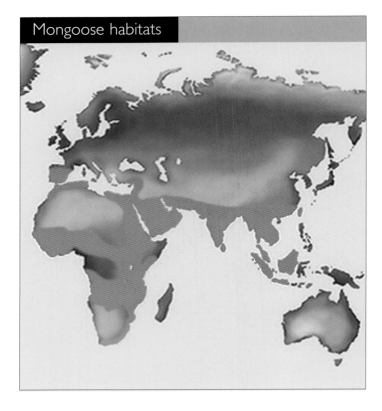

Mongoose habitats

many traits with cats, although they tend to have shorter legs and longer muzzles. The muzzle is the part of the face that includes the jaw and nose. In a cat, the muzzle is rounded, but, as a mongoose has more teeth, its jaw has been extended to accommodate them.

Several types of mongoose are found throughout Africa, southern Europe and Asia. The largest Asian variety is the Crab-Eating Mongoose, which lives mainly in the marshy valleys of the Himalayas. These acrobatic mammals can

Comparisons

The Slender Mongoose is found throughout much of sub-Saharan Africa. It tends to be one of the smaller of the many species of mongoose, although it is not the smallest. That distinction goes to the Dwarf Mongoose, which is only around 24cm (9in) long. Larger sub-species such as the Common Indian and the African Grey grow to between 40-60cm (16–24in).

African Grey Mongoose

Slender Mongoose

grow to 60cm (24in) on a diet of crabs, frogs and fish. Two other sub-species are found throughout Asia: the Small Indian Mongoose and the Common Indian Mongoose. The Common Indian Mongoose is probably the most familiar of the smaller varieties, with stiff yellowish-brown fur, dark paws and pink spots around the nose and eyes. Despite these exterior differences, all species of mongoose are naturally fast and alert, a prerequisite for an animal with such a death-defying lifestyle.

Battle Tactics

Snakes such as Black Mambas have been timed at speeds of up to 11km/h (almost 7mph) and even slower varieties can move with surprising celerity. The secret of the mongoose's success against such seemingly overwhelming odds is its own lightning fast reflexes and – in true military style – a thorough knowledge of its enemies' weaknesses. Mongooses use different strategies, depending on what type of snake they're hunting, but perhaps the most spectacular displays of daring and gymnastics are seen during their encounters with cobras.

When agitated, a cobra will rear up in a striking position. The mongoose is fast enough to anticipate the snake's slightest movements, and manages to keep the cobra agitated but just out of striking range, by jumping back and forwards in its line of sight. The cobra is eventually worn down by the effort required to keep its body erect. Once this happens, the mongoose moves in to deliver a killing bite on the back of the snake's neck. In this deadly game, mongooses aren't always the victors, but luckily nature has also given them a natural tolerance to snake venom.

This African Grey Mongoose spies a pied crow and puts on a crazed display, chasing its own tail to catch the bird's attention.

The curious crow flies down to get a better look at the mongoose as it tumbles around on the ground.

The inquisitive crow lands and pecks at the peculiar animal. It has foolishly brought itself within range of the mongoose.

Suddenly, the mongoose turns and strikes. It is already too late for the crow to escape.

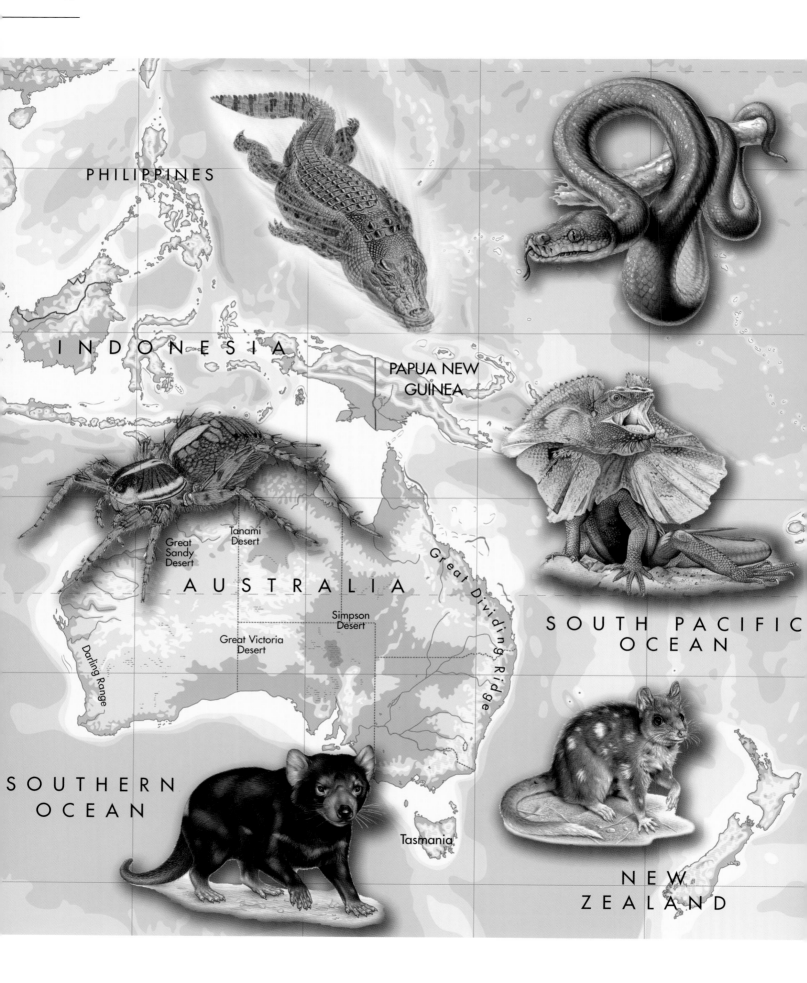

PHILIPPINES

INDONESIA

PAPUA NEW
GUINEA

Great
Sandy
Desert

Tanami
Desert

AUSTRALIA

Simpson
Desert

Great Victoria
Desert

Great Dividing Ridge

Darling Range

SOUTH PACIFIC
OCEAN

SOUTHERN
OCEAN

Tasmania

NEW
ZEALAND

Australasia

Australia is one of the true natural wonders of the world, being the only country that is also a continent. Combined with New Zealand and its neighbouring islands, the vast landmass that makes up 'Australasia' is famous for its wild, spectacular scenery and unique wildlife.

~

Scientists believe that the continents of the world once formed a huge, united landmass, which has been called Pangaea. This 'super continent' began to split apart around 200 million years ago. One half formed Laurasia, which eventually split again to create Europe, North America and Asia. The other half, Gondwanaland, ultimately became Africa, Antarctica, Australia and South America. It was during this great schism that Australia and its neighbouring islands became isolated from the rest of the world – preserving its unique flora and fauna until the arrival of the first European settlers in 1788. Despite the eradication of many native species, as farmers and colonists cleared the land, the region still boasts animal inhabitants that can be found nowhere else on earth.

Australasia has long had a reputation as home to poisonous and dangerous animals: of Australia's 140 species of snakes, for example, the Taipan and Tiger Snake are the deadliest in the world. Yet, if we were to take a moment to step into its blossoming rainforests or cross its scorching, sun-baked deserts, we would discover an Australasia that is much more dynamic and complex, and where danger doesn't always come with fangs attached!

Quoll

Before European animals were introduced to Australia, Spotted-Tailed Quolls ranked as one of the island continent's largest meat-eating mammals. Fierce, predatory and proficient hunters, quolls will attempt to make a meal of almost anything. In their own ecosystem, these ferocious little hunters are aptly nicknamed the 'Tiger Cat'.

Key Facts	ORDER *Marsupialia* / FAMILY *Dasyuridae* GENUS & SPECIES *Dasyurus*
Weight	0.3–7kg (10oz–15lb 6oz)
Length Head & Body Tail	12–75cm (4^3/4–25^1/2in) 12–55cm (4^3/4–21^1/2in)
Sexual maturity	1 year
Mating season	April to August, depending on species
Gestation period	12–21 days
Number of young	1 to 8 (up to 30 in the eastern quoll, but only 6 to 8 survive)
Typical diet,	Small mammals, birds, reptiles, amphibians, insects, spiders, earthworms, fruit, carrion and refuse in urban areas
Life Span	Eastern Quoll 3–4 years; other species unknown

Toes
Like the Tasmanian Devil, the Eastern Quoll is an unusual marsupial because each of its hind feet has only 4 toes; 5 is usual. Such an anomaly doesn't prevent this small marsupial from being an excellent climber.

Teeth
Quolls are 'polyprotodont'. This means that they have more than 4 incisors (the sharp, cutting teeth) in their lower jaw.

Quoll habitats

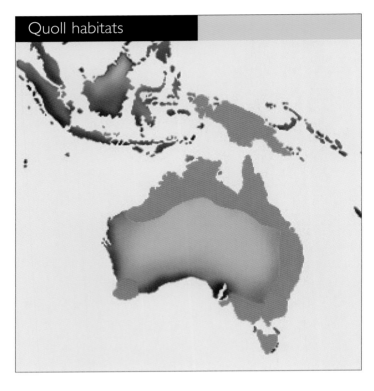

Quolls are among Australasia's best-known and well-loved animals. Yet many varieties are facing the threat of extinction because their habitats are being destroyed to make way for city living.

Second Best?

Marsupials are an early form of mammal that made their first appearance on earth around the same time as the dinosaurs. Modern mammals are usually placental: their unborn young grow inside the body, where they are supplied with nutrients and oxygen by an organ developed for that purpose, called the placenta. When the fully formed young are born, the placenta is expelled as afterbirth. In marsupials, the young are born at an extremely immature stage of development. New-born quolls, for example, are about the size of a grain of rice. To survive, these tiny, blind and almost limbless babies must crawl through their mother's fur until they reach a flap of skin called the marsupium (pouch). Once inside, the offspring attaches itself to a nipple, which provides it with milk. In the case of quolls, there are only six nipples for thirty babies, so only the first half-dozen babies will live. A young quoll stays attached to the nipple for eight weeks until it finishes developing and is ready to leave the pouch. This may take six to seven months for larger marsupials like kangaroos. Worldwide, placental mammals became the dominant 'design' because their young were more likely to survive to adulthood. However, because Australasia was isolated from the rest of the world for 200 million years, marsupials thrived without competition from placental mammals.

The Big and Small of It

In the wild, quolls are found in Australia and New Guinea only. There are four varieties of this small, furry native in Australia: the Spotted-Tailed Quoll, Eastern Quoll, Western Quoll and Northern Quoll. Quolls are also known as Native Cats or Marsupial Cats because of their similarity to felines. The eastern Australian Spotted-Tailed Quoll is the largest member of a species that varies in size

Comparisons

The parched, dry deserts of South Western Australia is where the Kowari makes its home. This fearsome hunter is smaller than the Eastern Quoll, but regularly kills animals as big, if not bigger, than itself. As food can be scarce in such extreme conditions, the Kowari is able to spend short-periods of time in a state of semi-hibernation, which is called 'torpor'. This handy skill allows the Kowari to save both energy and food reserves.

Eastern Quoll

Kowari

from about 75cm to 120cm (2ft 5in–3ft 11in). The smallest variety is the Northern, which is found in the tropical north. Quolls spend much of their time on the ground but are expert climbers: rough pads on their front paws help to grip the tree bark, and long claws offer added purchase.

Quolls live mainly in forested areas. The larger varieties will often make their den in a cave or hollowed-out log. The smaller ones may dig burrows. Like many predators, quolls are nocturnal hunters, using their burrows to sleep in during the day. When hunting, quolls are efficient and seemingly tireless, constantly moving from the ground to trees in search of a meal.

A Little Bit of Everything

Nature loves an opportunist. Quolls may be carnivores, strictly speaking, but their teeth tell another story. Like all carnivores, quolls have long, sharp canine teeth, which are used to tear apart flesh. Yet they also have flat molar teeth, which grind and pulp vegetable matter. This means that they're equipped to take advantage of any available food source, and they certainly do. A quoll's natural prey includes frogs, birds and lizards. The larger varieties will also tackle possums, which can weigh as much as a female quoll, as well as domestic birds like chickens. They also regularly supplement their diet with carrion, plant seeds and fruit. This flexible approach to food meant that quolls were one of Australia's success stories. Then came immigration, unfortunately, and as so often, this undermined evolution. When new settlers arrived in the eighteenth century, they brought with them many Europeans animals, including foxes and cats. Against these, the unsuspecting quoll simply cannot compete.

During the day, the quoll rests in its lair, but as night approaches it sets out to hunt for food.

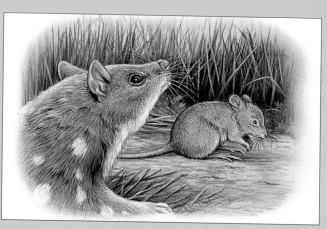

The quoll homes in on the sound of a nearby potoroo, a rabbit-sized marsupial.

Despite a surprise attack, the potoroo turns and defends itself. However, the quoll delivers a mortal bite to its neck.

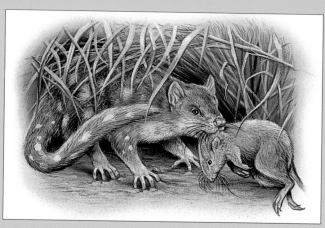

The quoll retreats into the undergrowth with its prize, away from the prying eyes of larger scavangers.

Green Tree Python

Among the giant trees of northeast Australia's flourishing rainforests, the Green Tree Python is master of all it surveys. This sinewy, silent snake may not be the largest of the pythons, but its spectacular natural camouflage makes it one of the rainforest's great, clandestine killers.

Juveniles

When they hatch, baby Green Pythons are more likely to be yellow, red or brown, than green. They slowly turn green over a period of about 6 months.

Key Facts	ORDER *Squamata* / FAMILY *Boidae* / GENUS & SPECIES *Morelia viridis*
Length	Up to 2m (6ft 6in), but usually about half this length
Weight	Up to 9kg (19lb 13oz), but usually 3–6kg (6lb 9oz–13lb 3oz)
Sexual maturity	2 years
Breeding season	End of wet season
Number of Eggs	10 to 20
Hatching period	About 60 days
Breeding interval	1 year
Typical diet	Mainly birds; also bats and other small mammals
Lifespan	Up to 35 years in captivity

Teeth

A python's teeth are backward-facing. This allows the snake to get a better grip on its prey.

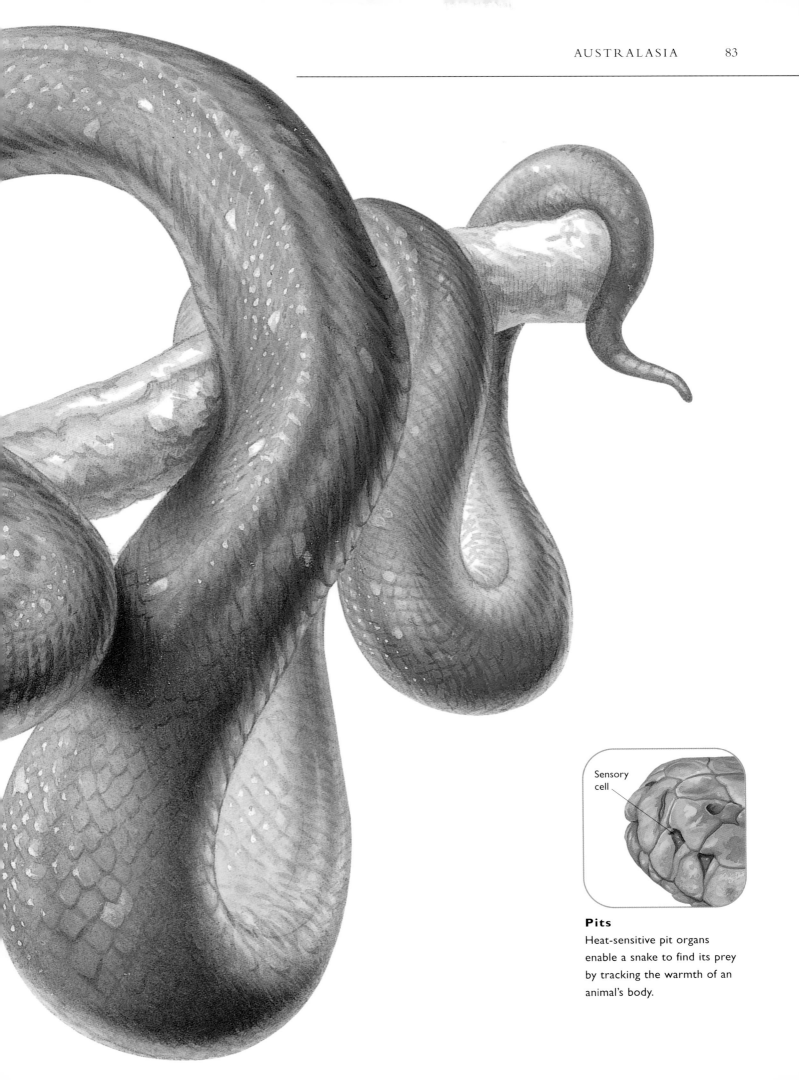

Pits

Heat-sensitive pit organs
enable a snake to find its prey
by tracking the warmth of an
animal's body.

Looking for its next meal, the Green Tree Python investigates an old hollow tree, gripping a branch to steady itself.

The python peers into the hollow, using infrared sensors situated on its upper lip to search the darkness for prey.

The python selects a victim from one of the colony of bats roosting inside the tree and prepares to strike its target.

The sleeping bat is helpless as the snake, unable to chew, expands its elastic jaws and swallows its prey whole.

Pythons are among the most widespread and successful members of the snake family. The largest and best known are the African and Asian, but the slightly smaller Green Tree Pythons are equally diverse. Varieties can be found throughout the world's tropical regions, including areas of Iran, New Guinea and Australia.

A World of Colour

Despite its name, the Green Tree Python can be quite variable in colour. In fact, newly hatched baby Green Pythons are more likely to be yellow, red or brown than green. After about six months, they slowly begin to change colour. This transition may happen within a week, but it can take up to three months for the python to attain its final adult coloration. Even then, it might not be green; it

could be yellow or blue. Blue varieties are quite rare and especially valuable to snake breeders.

Many animals have the ability to change their coloration. The chameleon is an extreme example of this phenomenon: this amazing lizard can alter its skin colour in response to its surroundings, using hormones in its body, which react to changes in mood, temperature and light. So, when it's threatened, it can become almost invisible. Yet even birds change their plumage from winter to summer, allowing them to better blend in with the changing environment. Herpetologists have not yet discovered the reason why Tree Pythons change their colour as they grow, but it's likely to be protective, as young pythons, like may juvenile animals, are at risk from a wide range of predators.

Comparisons

In order to survive in different environments, animals must adapt. When conditions are the same, animals will develop similar survival traits. This is called parallel evolution, and a good example of this can be seen in the Green Tree Python and the Emerald Tree Boa. The Emerald Tree Boa, like the Green Tree Python, lives in the rainforest, although the boa is a native of South America. Their natural habitats are so alike, however, that they have developed to look and behave in a strikingly similar way.

Emerald Tree Boa

Green Tree Python

Now You See Me …

For many animals, effective camouflage is not just defensive. It plays an important role in hunting too. The spots on a cheetah's coat help to hide it when it's stalking prey in the long grasses of the African savannah. The high contrast between the black spots and pale background break up the cheetah's body shape and, from a distance, make it hard to see. This type of camouflage, called disruptive coloration, has been copied for centuries by humans in times of warfare. The adult Green Tree Python uses a form of camouflage called cryptic coloration, which enables it to blend in with its surroundings. Although their colours may seem startlingly vivid to us, it is perfect for its environment, as Green Tree Pythons are totally 'arboreal', sleeping, breeding and, probably, nesting among the trees. It's here too that they hunt, using techniques that utilize their natural camouflage to greatest effect.

Now You Don't

Tree Pythons need to eat only every 10 to 14 days. This is slightly more often than most pythons, and is necessary because nesting females often go without food for the entire five-month brooding period and so need to build up fat reserves. Not having to eat every day has advantages – pythons can afford to be patient predators. Wrapping its long body around the branches of a tree, it can sit, virtually motionless, for days, until a meal passes its way. Once in position, it is so still and silent that prey may even crawl over its coiled body, unaware that death is just inches away.

During the hunt, the Green Tree Python will typically use its prehensile tail to anchor its body to the tree, lowering down the remaining third of its mass in search of a meal. Once within striking distance, the python will sink its teeth into its victim. Birds form a large part of a tree python's diet and, like many bird-eating snakes, it has especially long teeth that can penetrate feathers and anchor this 'fast food' in one place. In a similar way to Reticulated Pythons, they then suffocate their victim, constricting and eating it while hanging head down.

Green Tree Python habitats

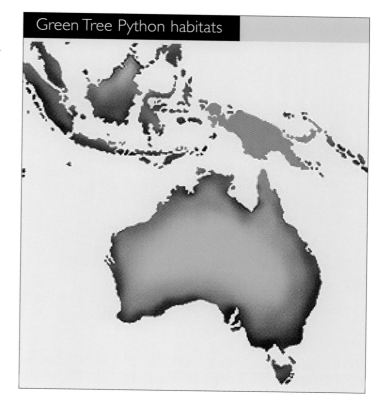

Frilled Lizard

In the animal kingdom, appearances are everything. Sometimes, appearing to be dangerous is one of the most effective ways of protecting yourself against larger, deadlier predators; the Frilled Lizard is one of nature's great pretenders.

Key Facts	ORDER *Squamata* / FAMILY *Agamidae* GENUS & SPECIES *Chlamydosaurus kingii*
Weight	Up to 740g (1lb 10oz)
Length Head & Body Tail	25cm (10in) 51cm (20in)
Sexual maturity	2–3 years
Breeding season	Spring
Number of eggs	4 to 13, average 8
Incubation period	175–95 days (at 30°C)
Birth interval	More than one clutch may be laid in each breeding season
Typical diet	Chiefly moth and butterfly larvae, termites and ants; also small mammals
Lifespan	Up to about 5 years in the wild

Teeth
At the front of the Frilled Lizard's mouth are sharp, piercing teeth. To the rear are smaller cutting teeth. Unlike human teeth, which sit in sockets, these rear teeth are actually part of the jaw.

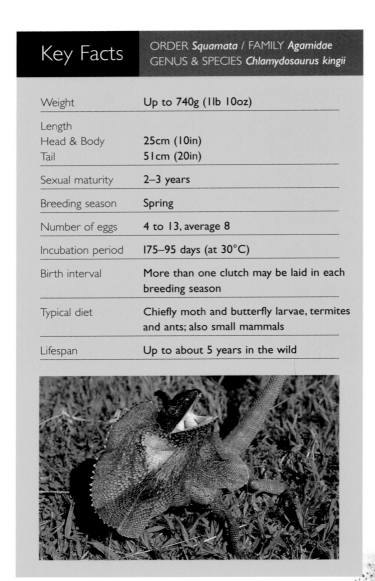

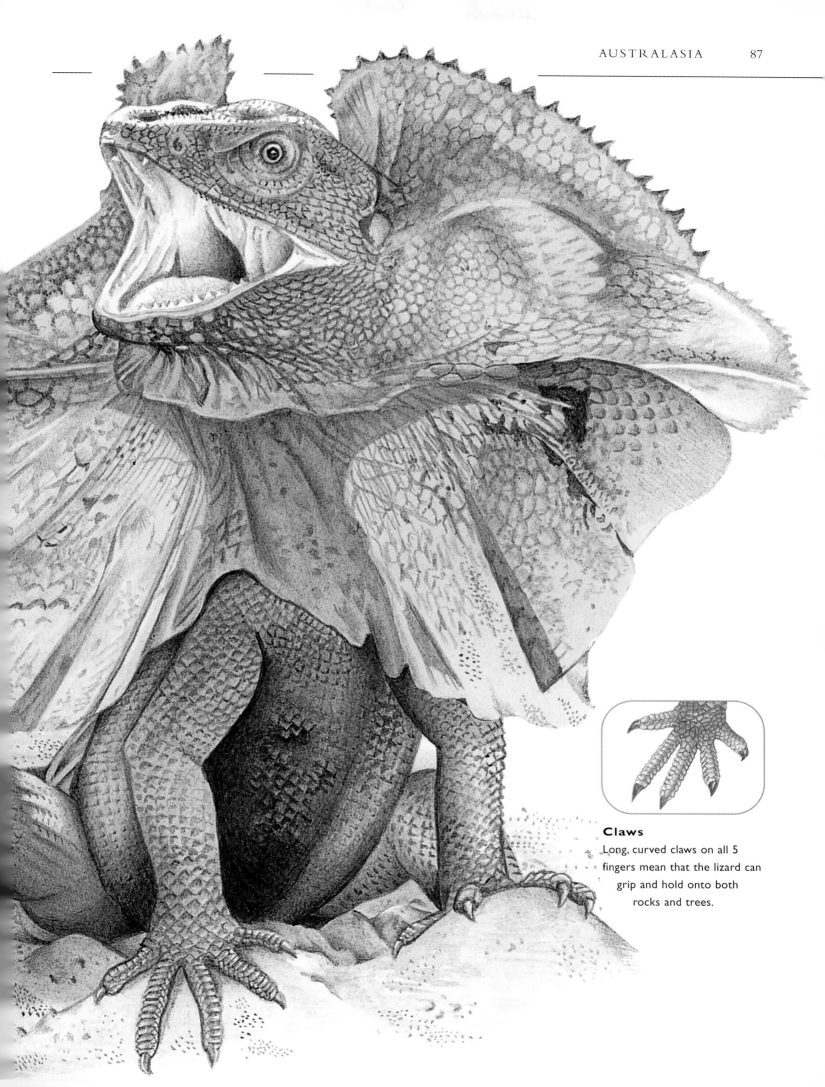

Claws
Long, curved claws on all 5 fingers mean that the lizard can grip and hold onto both rocks and trees.

The Frilled Lizard belongs to the family *Agamidae*, which includes around 300 species common to the warmer parts of Africa, Asia and southern Europe. In Australia, this slender reptile makes its home in the sandy, semi-dry regions of the north and northeast.

Misleading Messages

Throughout the world, animals send signals to each other. The highly poisonous Harlequin Coral Snake, for example, uses its orange and black warning stripes to tell would-be predators, 'I'm dangerous: stay away.' The Bumble Bee gives out a similar message, in yellow and black. The brightly patterned Monarch Butterfly has a bitter taste. Over thousands of years, birds have come to associate this taste with the butterfly's coloration, so they no longer eat them.

The valuable lesson that many animals have learnt is that they don't necessarily need to have poison, stingers or a bad taste. They just need their enemies to think they do. The False Coral Snake, for example, has similar markings to the Harlequin, but is totally harmless. The Flower Fly looks like a Bumble Bee, but has no stinger. The Viceroy Butterfly has a similar coloration to the Monarch but, compared to its counterpart, makes a tasty snack. This form of protective coloration is called Batesian Mimicry, named after the English naturalist Henry Bates (1825–1892), who first identified it.

Frills and Thrills

The Frilled Lizard's defence mechanism isn't based on mimicking any specific animal, but on making itself appear much larger and more dangerous than it really is. This is

As a predator approaches, the lizard remains still and keeps its frill flat, hoping that it will go unnoticed.

If the predator continues to approach, the lizard suddenly spreads its frill and opens its mouth, hissing. The display of yellow and pink is designed to frighten enemies.

The lizard will then rear up and attack its persistent attacker with sharp teeth and whiplike tail.

Finally, if the assailant cannot be driven off, the lizard will sprint away or climb the nearest tree to escape.

Comparisons

Just off the northwestern coast of Australia lies the island of Komodo, one of the 13,500 or so isles that make up the country of Indonesia. Yet it has become famous as the home of the world's largest lizard, the Komodo Dragon. When compared to the lowly Frilled Lizard, the Komodo is even more impressive, at around 200 times its weight and 5 times its length.

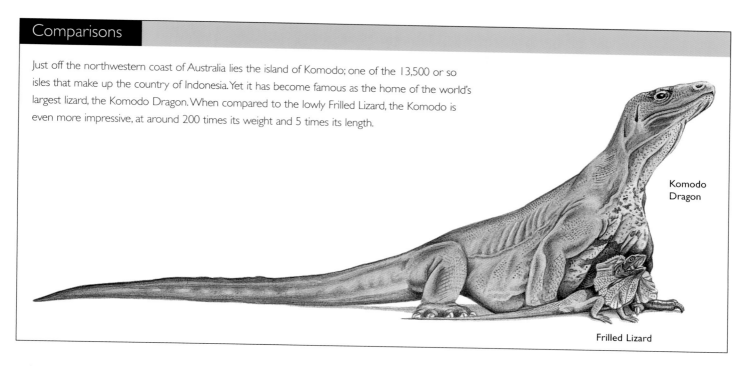

Komodo Dragon

Frilled Lizard

achieved using the large flap of skin around its neck. Normally, this flap lies over the lizard's shoulders, but in times of danger it flips open, like an umbrella, using tough, elastic tissue called cartilage to hold it erect. Fully raised, this frill can measure 20cm (8in) in diameter, which is as long as the lizard's head and body combined. To add to the intimidating effect, the lizard will also open its mouth and sway from side to side to give the impression that its head is even bigger. If this initial warning is ignored, the lizard will begin to walk towards the intruder, hissing loudly. Despite the fact that the Frilled Lizard normally eats nothing larger than spiders, eggs and the occasional bird, this bluff usually pays off. Even Australian wild dogs, which regularly attack larger, more dangerous animals, will quickly back away from this apparently fearsome predator.

An Actor's Life ...

Frilled Lizards are designed for forest living. Their long, skinny forelegs and large hind legs make it easy for them to grip and climb trees. Especially at night, the lizard is most at home high in the branches, where it is relatively safe from predators, such as dingoes and quolls. Frilled Lizards are solitary by nature, but during the mating season in September they go to elaborate lengths to attract a mate. During this time, males become extremely aggressive to other males and often fight over territory or females. To attract a female, the male will perform a courtship dance. If she's interested, she'll respond by bobbing her head. Once mating has occurred, the female will lay between eight and twenty tiny eggs. It takes around eight to twelve weeks for the young lizards to hatch, but they're fully independent

once they do – and ready to begin their own fight for survival. Luckily, on the occasions when the lizard's posturing fails, it's also a fast runner. By raising the front of its body, the Frilled Lizard can run at great speeds on just its back legs. Interestingly, the tracks it leaves in this way are similar to those made by dinosaurs of the Mesozoic Era, 240–63 million years ago. Clearly, this great bluffer has been getting away with this act for quite some time!

Frilled Lizard habitats

Leopard Seal

Leopard Seals were given their name because of their spotted coats, but have a reputation that their land-bound namesakes would envy. Aggressive hunters with voracious appetites, Leopard Seals are one of the top predators in the southern hemisphere.

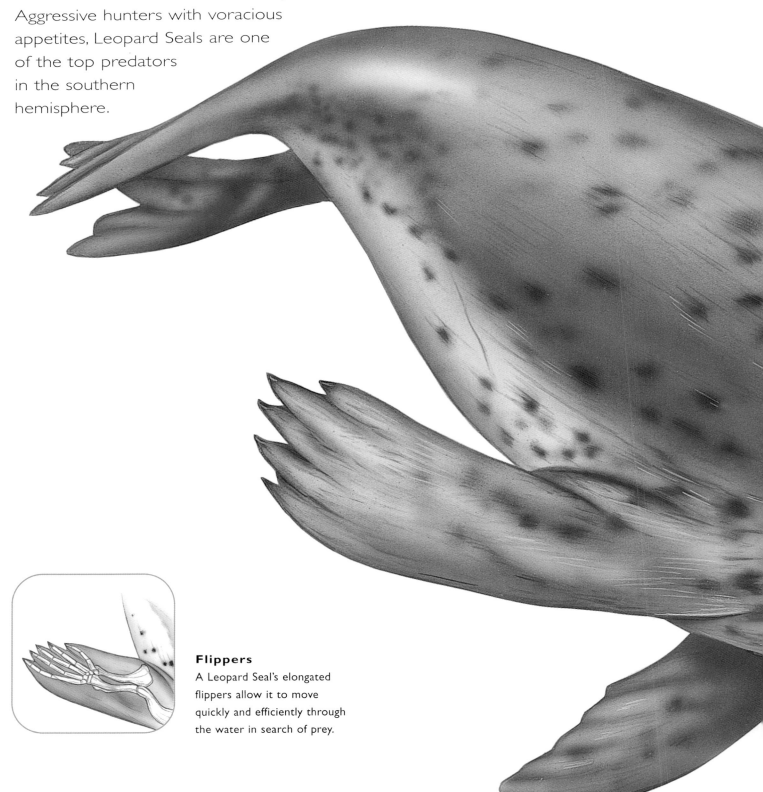

Flippers
A Leopard Seal's elongated flippers allow it to move quickly and efficiently through the water in search of prey.

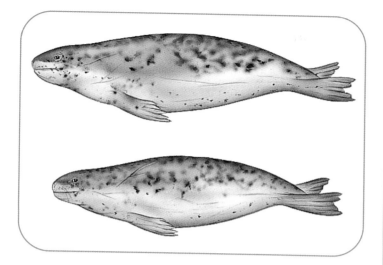

Male and Female

In much of the animal kingdom, males tend to be bigger than females. Female Leopard Seals are, however, around 5 percent bigger than their male counterparts.

Key Facts	ORDER *Pinnipedia* / FAMILY *Phocidae* / GENUS & SPECIES *Hydrurga leptonyx*
Weight	Male 325kg (717lb); female 370kg (816lb)
Length	Male 2.8m (9ft 2in); female 3m (9ft 10in)
Sexual maturity	Male 4 years; female 3 years
Mating season	November to February
Gestation period	9 months, plus 3 months delayed implantation
Number of young	1
Birth Interval	1–2 years
Typical diet	Krill, fish, squid, penguins and other seabirds, young seals
Lifespan	About 25 years

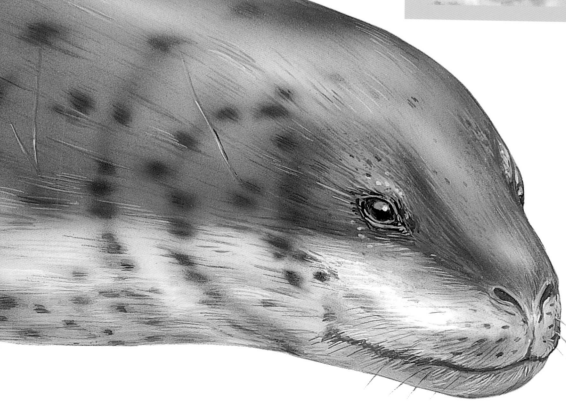

Teeth

A staple of the Leopard Seal's diet is krill. They eat this by straining through their teeth, which have become specially adapted for the job.

Leopard Seal habitats

Leopard Seals are found widely throughout the southern half of the globe. They make their homes mainly on the edges of the Antarctic ice packs, where they mate and reproduce, but they are also among Australasia's more unexpected visitors, appearing regularly during winter and spring along the coastline of New Zealand, Australia and nearby islands.

Hunter-Killers

Leopard Seals are the largest and most hostile members of the seal family (family *Phocidae*). It has been said that

Leopard Seals will attack and try to eat almost anything, but the bulk of their diet is probably made up of krill, which is a small shrimplike animal, eaten in vast quantities by whales. Octopus and squid are also high on the menu. Yet these giant mammals didn't get their reputation as killers by eating such relative 'small-fry'.

It's in their energetic pursuit of penguins that a Leopard Seal's hunting skills can be seen at their best. In less than a minute, a Leopard Seal can leap out of the water, grab a penguin straight from the ice and smash it on the water's surface. They do this with such force that the resulting boom as the bird's body is slammed against the water can be heard more than 1 kilometre (1100 yards) away. A Leopard Seal may catch and eat up to six penguins in this way in just over an hour. Leopard Seals also regularly make a meal of other members of the family *Phocidae*, particularly Crab-Eater Seals. In fact, they're the only seals that feed on warm-blooded animals, and this list includes some surprising delicacies. One seal caught on the Australian coast, for example, had a Duckbill Platypus in its stomach. Unlike other seals, Leopard Seals have long, sharp teeth – ideal for biting – and they've even been known to try them out on unfortunate scientists or tourists who've strayed too close!

From Pup to Predator

Leopard Seals are solitary animals and can be found in groups only during the breeding season, from mid-

Comparisons

One of the more familiar relatives of the Leopard Seal is the Antarctic Fur Seal, so called because it is hunted for its fur. Fur Seals, Sea Lions and walruses all have the ability to turn their flippers downwards. This means that they can walk, effectively, if slowly, on all fours. The Leopard Seal, doesn't have this ability. Instead, when on land, it propels itself forwards by lying flat on the ice and constricting the muscles in its stomach, to create forward motion.

Antarctic Fur Seal

Leopard Seal

November to December. Mating takes place in the water, and a single baby seal (pup) is born after an 11-month pregnancy. Males are rarely seen near the 'rookeries', and they probably continue their separate lives fairly soon after the birth. For the female, though, the job of teaching the young Leopard Seal the skills it needs to survive has just begun. At birth, the pups may weigh as little as 22kg (50lb). They look very like miniature adult seals except for their soft, thick coat, which they must shed before they can learn to swim. Like adults, the pups have healthy appetites and quickly progress from krill to larger prey, growing to adulthood at around three to six years old.

At Home, at Sea

Leopard Seals have a streamlined design, made to measure for a life at sea. Their long bodies are slender so that they can move through the water efficiently. Elongated front and back flippers are used to propel them forward, and help them to change direction. These adaptations enable a seal to swim at a top speed of 16km/h (almost 10mph). On land, Leopard Seals are much less mobile. All seals have a layer of thick blubber (fatty tissue), which helps them to keep warm. It also makes them big, bulky and slow on land. Fortunately, Leopard Seals are so ferocious that they have few natural enemies, apart from the Killer Whale.

The Leopard Seal watches the penguins from beneath the water, patiently waiting for them to enter the sea.

Unaware that the seal is lying in wait, the penguins dive in. The seal surges forward and strikes.

The mouth of the seal is lined with long, curved teeth, which enable it to seize the penguin and hold it.

The seal is unable to swallow the penguin whole, so it removes the skin by vigorous shaking. The flesh is then eaten but some parts, such as the head and bones, are discarded.

Saltwater Crocodile

The Saltwater Crocodile is the world's largest living reptile. Aggressive and highly territorial, this saw-toothed hulk has such a bad reputation that it generates fear wherever it makes its home.

Palate valve

Even when its mouth is shut, water is able to seep in through the crocodile's jaws. To prevent itself drowning, a special valve in its throat seals its mouth underwater.

Skull

Eyes and nostrils, on the top of their skull, allow crocodiles to sit almost completely submerged in the water until prey approaches. Once a meal is in sight, powerful jaws close with tremendous force, making it difficult for victims to escape their grasp.

Key Facts	ORDER *Crocodylia* / FAMILY *Crocodylidae* / GENUS & SPECIES *Crocodylus porosus*
Weight	Male average 500kg (1102lb); female smaller
Length	Male average 4.5m (14ft 9in), maximum 7m (22ft 10in); female a third smaller
Sexual maturity	Male at 3.2m (10ft 6in) or 14–16 years; female 2.3m (7ft 6in) or 10–12 years
Breeding season	Usually start of rainy season
Number of eggs	25 to 90, usually 40 to 60
Incubation period	90 days
Breeding interval	1 year
Typical diet,	Mammals, birds, fish, reptiles, amphibians and crustaceans
Lifespan	Average 40–50 years, but up to 100 or more

Hindfeet

Crocodiles use their webbed feet, in conjunction with a muscular tail, to propel themselves through the water.

Saltwater Crocodile habitats

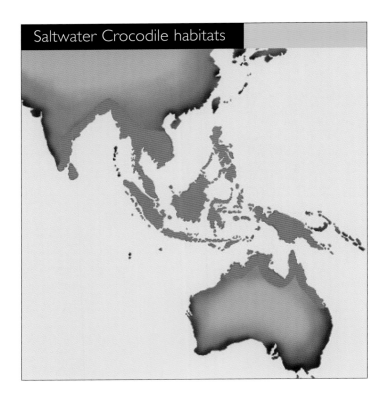

If we were able to travel back in time 200 million years, what would we see? Well, of all the weird and wonderful creatures roaming the earth, the most familiar would be the crocodilians. This ancient and widespread group of reptiles has been so successful as a species that they've evolved very little since the Mesozoic Era.

A Distinctive Species

There are three distinct crocodilian families: alligators and caimans, gavials, and true crocodiles, of which the Saltwater Crocodile is the largest example. Although

similar to large lizards in appearance, crocodilians have many unusual features. In common with amphibians, for example, they have webbed feet. They also have some distinctly mammalian attributes. These include an extremely efficient four-chambered heart and teeth that are set in sockets in the jaw, rather than fixed to the bone. Saltwater Crocodiles are especially unusual as most crocodiles prefer freshwater habitats. The Saltwater Crocodile is, in fact, the most marine of all these giant predators. In Australia, they tend to be found in estuaries and brackish (briny) coastal water. Some have even been found as far as nearly 1000km (600 miles) out to sea. However, they do migrate between habitats, and lower status males are often pushed out into freshwater rivers, swamps and billabongs (stagnant pools) by dominant males protecting their territory and breeding rights.

A Crushing Grip

Crocodiles are well equipped for the life of a hunter, with huge, incredibly powerful jaws. It is estimated that a 1-tonne (1.1-ton) crocodile can exert a crushing force equivalent to 13 times its own weight. One lucky zoo keeper in Australia experienced this vicelike grip at first hand and lived to tell the tale.

When removing the large male from its enclosure, the usual technique was to place a loop of rope around the crocodile's snout to keep its jaws closed. As long as the keeper pulled forward, the crocodile would naturally resist and pull back, keeping the loop taught. However, when the keeper momentarily loosened his grip on the restraining rope, the crocodile shot forward and grabbed his arm. The croc then began its 'death roll', which is designed to stun and dismember large prey. Still holding

Comparisons

While the Saltwater Crocodile prefers coastal regions, Australia's freshwater rivers, lakes and swamplands are home to an equally impressive predator. The Freshwater Crocodile – nicknamed 'freshie' by locals – is a relative of the Saltwater Crocodile, but tends to be much smaller. Living almost exclusively on a diet of aquatic animals, freshie is still a force to be

reckoned with. While it is less aggressive than the saltwater variety, it has still been known to attack human bathers during periods when its natural food is in short supply.

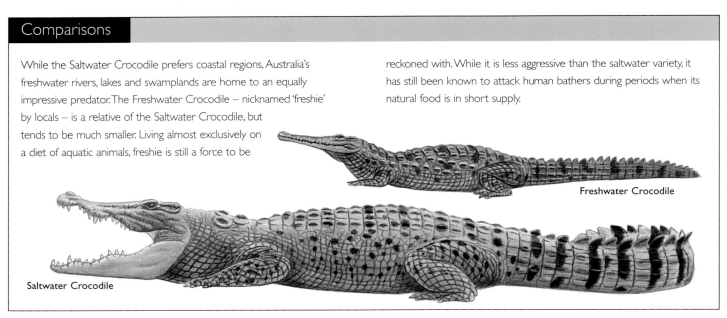

Freshwater Crocodile

Saltwater Crocodile

the keeper's arm, the huge reptile began rolling its body over and over. Despite his agony, the keeper knew that, if he didn't tumble with the crocodile, his arm would be torn off. After just 20 seconds, man and crocodile were separated, but this was enough time for the bones and nerves in the keeper's arm to be totally shattered.

On the Menu

Saltwater Crocodiles eat a variety of prey, including large mammals such as humans. There are many well documented horror stories about shipwrecked mariners being attacked by Saltwater Crocodiles during World War II. Even today, people in this region are rightly cautious in their dealings with these renowned killers. Fortunately, the Saltwater Crocodile's preferred food is less close to home. Juvenile crocs tend to feast on smaller prey, which may be nothing larger than insects for the first few years of their lives. As the crocodile develops, so too do its hunting skills. Young crocodiles are very buoyant, so they have to swallow stones to help them lie low enough in the water to stalk their prey effectively. With practice, though, they quickly progress to a diet of frogs, lizards, mud crabs and turtles. As they grow larger, hunting becomes easier. A blow from the tail of a 7m (23ft) crocodile is enough to stun most animals. Once disabled, prey is dragged underwater and drowned. Saltwater Crocodile's stomachs have extremely large muscles that help to grind flesh, so digesting a large meal like a buffalo is no problem.

Soon after mating, the female constructs a nest mound up to 1m (3ft 3in) high and 2.5m (8ft 3in) across near the waterside.

During the 90-day incubation period, the mound is fiercely defended from predators such as monitor lizards.

When the eggs begin to hatch, the mother is alerted by stirrings in the mound and digs out her babies.

The mother gently carries the newborns to the water in her powerful jaws so they can swim for the first time.

Tasmanian Devil

The Tasmanian Devil is Australasia's largest carnivorous marsupial. With its ferocious appearance, territorial nature and blood-curdling howls, a fully grown 80cm (30in) 'devil' is well named.

Feet
The Tasmanian Devil is unusual for a marsupial in that each of its front feet has 5 toes and its back feet have only 4. Each toe is tipped with a short claw.

Teeth

Within these powerful jaws are rows of large heavy molar teeth. These teeth are traditionally used for crushing and grinding food, and are ideal for breaking open bones.

Key Facts	ORDER *Marsupialia* / FAMILY *Dasyuridae* GENUS & SPECIES *Sarcophilus harrisii*
Weight	5–10kg (11–22lb)
Length Head & Body Tail	52–80cm (20–31in) 23–30cm (9–12in)
Sexual maturity	2 years
Mating season	March to April
Gestation period	31 days
Number of young	2 to 4
Birth interval	1 year
Typical diet	Mainly carrion; also live prey, ranging from insects to other marsupials
Lifespan	Up to 8 years

Tasmanian Devils were once common throughout Australasia, but they probably died out following the arrival of dingoes, which came to the continent with Aboriginal people about 50,000 years ago. Although there have been numerous sightings on the mainland, it is believed that Tasmanian Devils can now be found only on Tasmania, making them a unique and irreplaceable part of the island's ecosystem.

Teeth and Claws

It would be hard to mistake the Tasmanian Devil for any other animal. Although only the size of a European badger, this powerful marsupial both looks and sounds much more fearsome.

The devil's teeth and jaws are probably its most noticeable, offensive weaponry. A Tasmanian Devil can bite with the force of an animal four times its size. Young devils are agile hunters, but as they get older they rely more and more on carrion. Their heads get proportionally bigger, so that their jaws have more leverage to crush bones and break through skin. In very old animals, the head may contribute up to a quarter of its body weight.

Devils can open their jaws to 120° and often display their teeth in huge, gaping yawns. This yawn may be an example of 'displacement activity'. In times of stress, both humans and animals often do something that seems out of place: you may, for example, scratch your nose when you're nervous. This behaviour may be an attempt to show that, rather than being anxious, you're actually relaxed and confident. When a Tasmanian Devil shows you a mouthful of teeth, however, you probably don't want to stay around to find out what it means!

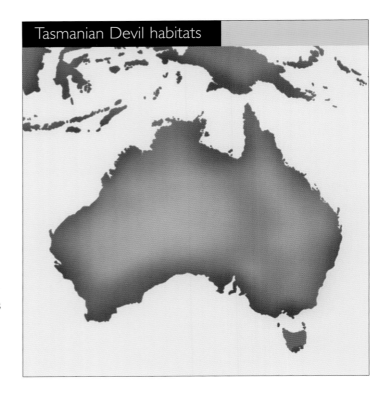
Tasmanian Devil habitats

Howls in the Night

Despite their intimidating appearance, Tasmanian Devils probably acquired their name because of the noises that they make. Devils are highly vocal animals and when they're fighting they let out spine-chilling howls and wails, which must have sounded distinctly eerie to early European settlers.

Most animal groupings are hierarchical. They have a 'pecking order' that determines which males or females will, for example, get the largest share of the kill, or

Comparisons

Eastern Quoll

Tasmanian Devil

The Tasmanian Devil is not Tasmania's only carnivorous marsupial. This wild little hunter shares its home with the much lighter, smaller quoll, which is often called the Native Cat. Australia has 4 varieties of this furry little predator: the Spotted-Tailed Quoll, Eastern Quoll, Western

Quoll and Northern Quoll. It is the Eastern Quoll that is the Tasmanian Devil's nearest neighbour, but both, unfortunately, are now quite rare.

control the best territory. This avoids unnecessary conflicts within the group. Devils aren't especially territorial, nor do they hunt in packs, but males will fight each other for food and access to females during the mating season. Female Tasmanian Devils often have scars around their neck, since the male will often bite the female on the neck to subdue her during mating.

All You Can Eat

One of the names for the Tasmanian Devil is Harris' meat lover, in honour of the man who first discovered them. As the name implies, devils are primarily carnivorous. They will eat almost any meat, from small insects and birds, to the occasional sheep, and even other Tasmanian Devils if they're hungry enough, although their favourite snack is wombat.

Like many predators, devils will often eat huge amounts of meat in a single sitting. They can consume around 40 per cent of their body weight in half an hour. Much of the reason for these eating binges is that hunting isn't always successful. A predator will always eat as much as it can when the opportunity arises, since it has no idea where or when the next meal is coming from. Even such large amounts of food, though, will last an average-sized devil for only a few days. Devils have long had a reputation as scavengers, rather than proper hunters, and it is true that they will eat carrion. Yet it is in the wilderness that Tasmanian Devils are at their most predatory and impressive. On rough terrain, a devil can run faster than a human. They are also excellent swimmers and able climbers, which means they can adapt to most environments, making them formidable hunters.

The Tasmanian Devil uses its acute sense of smell to hone in on carrion as it searches for its next meal.

The trail leads to the carcass of a dead wallaby. Sensing a good opportunity, the Tasmanian devil tries to eat as much as possible before competition arrives.

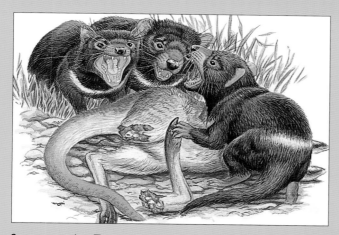

Soon two other Tasmanian Devils arrive, drawn by the scent of flesh and blood. They attempt to intimidate their rivals with growls and screams.

One Tasmanian Devil sees his chance to gorge himself while the others threaten one another with displays of strength, jaws gaping open.

Wolf Spider

Considering all the large, poisonous spiders that inhabit Australasia, it may seem strange to pick out the Wolf Spider as 'dangerous'. Yet, these large arachnids received their name because of their wolf-like hunting prowess. They are accomplished predators and can track and kill their victims without the aid of webs.

Key Facts

ORDER *Araneae* / FAMILY *Lycosidae* GENUS & SPECIES *Various*

Length	4–35mm ($^3/8$–$1^1/4$in)
Legspan	15–100mm ($^1/2$–4in)
Sexual maturity	9–15 months
Breeding season	Summer in temperate climates; all year round in tropical regions
Number of eggs	50 to 200
Incubation period	1–2 weeks
Breeding interval	Each female usually rears 2 to 4 sacs of eggs in her lifetime
Typical diet	Small insects and spiders
Lifespan	Up to 2 years

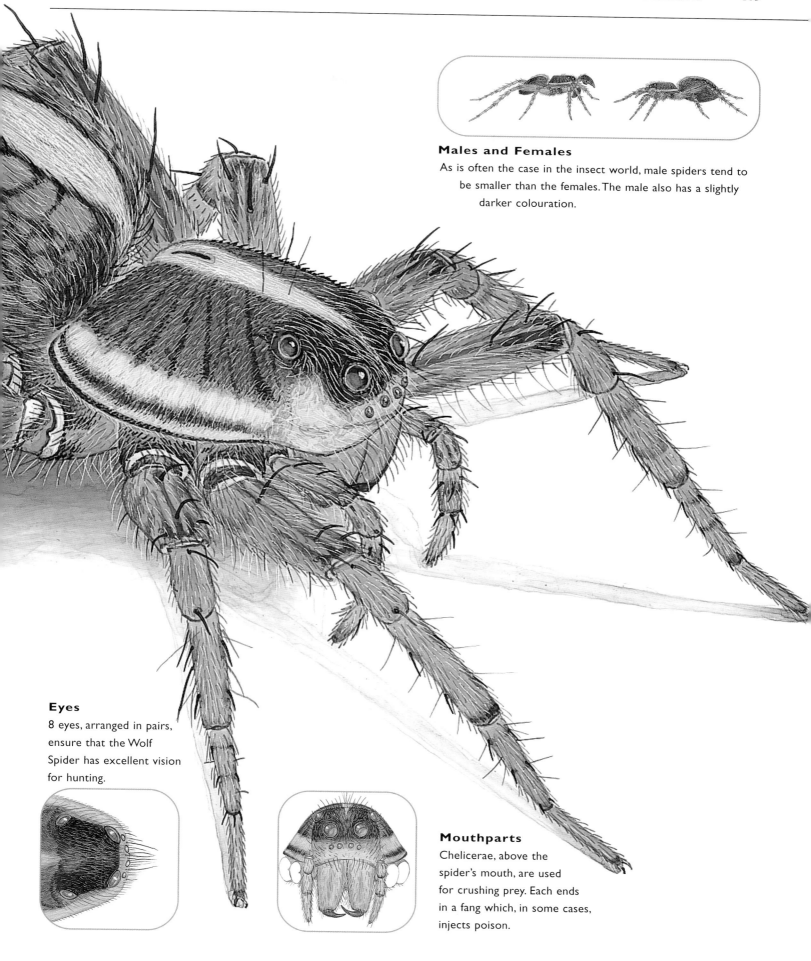

Males and Females

As is often the case in the insect world, male spiders tend to be smaller than the females. The male also has a slightly darker colouration.

Eyes

8 eyes, arranged in pairs, ensure that the Wolf Spider has excellent vision for hunting.

Mouthparts

Chelicerae, above the spider's mouth, are used for crushing prey. Each ends in a fang which, in some cases, injects poison.

Worldwide, there are more than 30,000 known varieties of spiders. This successful species can be found almost anywhere, including, amazingly, underwater and on the snow-topped peak of Mount Everest. Spiders belong to a group called arachnids. Unlike insects, which have six legs (three sets of two), arachnids have eight legs (four sets of two) and no wings.

True Hunters

All spiders have the ability to spin silk. Using spinnerets on their abdomen, many varieties use this silk to weave webs in which to trap their prey. Some, like the Bolas Spider, have a slightly different approach. They spin a line of silk with a knot at the end, which they swing at their prey, to entangle it, like a South American bolas. Hunting spiders, such as the Wolf Spider, also produce silk, but generally don't use it to make webs or trap prey. Instead, they stalk their prey and, when it's within reach, pounce. They use their silk for other, non-lethal activities, like lining nests. The female Wolf Spider spins sacs to protect her eggs, which she carries around attached to the spinnerets.

The reason that hunting spiders don't need to spin webs is that they have excellent eyesight. Wolf Spiders are diurnal: they're at their most active during the day, so well-developed vision is crucial to the hunt. They also tend to be larger than other spiders, with a powerful pair of chelicerae. Located just above the spider's mouth opening, these are used to crush prey. Each chelicera ends in a hollow fang, which, in some cases, is used to inject poison into the victim's body.

Amazing Variety

There are many types of Wolf Spiders. Those we're most familiar with, as occasional visitors to our homes and gardens, are probably large, with hairy bodies, but many have adapted to life in very specific environments. Some, for example, make their homes near water. These look and behave much like Fisher Spiders, which are lightweight spiders able to walk on water and dive for short periods of time. What all varieties of Wolf Spiders have in common is their stealth and speed. Depending on their habitat, Wolf Spiders may be grey, brown or black. This natural camouflage helps them to stalk their prey, which they then run to ground with impressive swiftness. The remarkable thing about such an athletic display is that spiders actually have no muscles to extend their legs. They can only move if their blood pressure is high enough!

Courtship and Parenthood

Large Wolf Spiders may live for several years, during which time they can have thousands of young spiderlings. As soon as the male matures, he seeks a mate. Since Wolf Spiders

Comparisons

Originally, the name tarantula referred to a type of Wolf Spider. Today, this name is used to refer to large, hairy spiders, of the type found in Mexico and South America. The Mexican Red-Kneed Spider (or tarantula) belongs to the same order of species as the Wolf Spider. Both are hunting spiders, who stalk and run down prey rather than build webs. Both are fast and aggressive predators. The main differences are their size and hunting techniques. The Mexican Red-Kneed Spider is considerably larger than the Wolf Spider – around 5 times the size. Red-kneed spiders also tend to hunt using their sensitive leg hairs to detect movements, while Wolf Spiders use their excellent eyesight.

Wolf Spider

Mexican Red-Kneed Spider

have such excellent eyesight, they use a particularly elaborate courtship dance to attract a partner. During the dance, the male raises his front legs and waves his pedipalps. These appendages look a little like antennae, but are actually positioned on the side of the mouth. A pedipalp is divided into six sections. The section closest to the mouth is used to cut and crush food, but the last section holds the spider's reproductive organs. So he's giving the female spider a clear message of his intentions!

Before mating, the male spins a sperm web, into which he deposits sperm. He then uses the pedipalps to transfer this sperm to the female, which she uses to fertilize her eggs – between 100 and 2000 at a time, depending on her size. Once hatched, Wolf Spiders carry the spiderlings on their back until they are ready to set out on their own.

Wolf Spider habitats

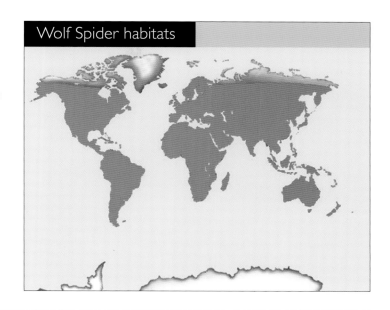

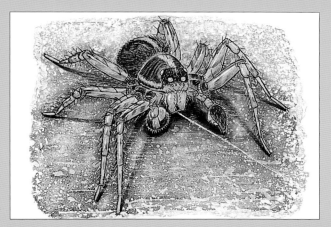

Following a silk thread that the female always trails, the male Wolf Spider sets off to find her.

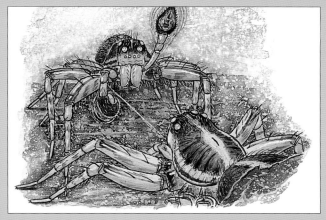

In an attempt to get her attention, the male 'dances', vibrating his legs and abdomen before waving his palps, his sex organs, in front of her.

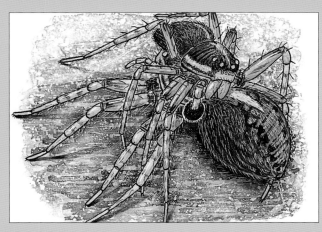

To see how keen the male spider is, the female may sometimes make fake attacks. The male, though, is resolute and strokes the her with his palps.

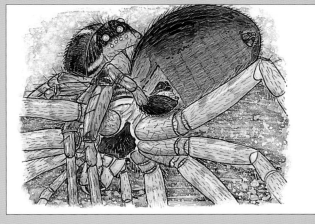

Eventually, the female allows the male to mount her. This allows him access to her underside, where her genital opening is situated.

ARCTIC OCEAN

Brooks Range

Mackenzie Mts.

ALASKA

CANADA

Ro

Mountains

Hudson
Bay

LABRADOR
SEA

UNITED
STATES

NORTH
PACIFIC
OCEAN

Sierra Nevada

Appalachian

NORTH
ATLANTIC
OCEAN

Gulf of
Mexico

North America

Everything about North America is big.
It may not be the largest of the world's great landmasses, but no other
region seems to be built on quite the same scale.

~

Technically, the North American continent includes Mexico and Central America, but the diverse wildlife of these southern regions will be dealt with in the South and Central America chapter. Even restricting our attention, however, to the huge upturned triangle of land that is North America and to the enormous island of Greenland takes us on a truly monumental journey: through the United States and up into Canada, from desert to polar ice sheets.

Starting this epic expedition close to the west coast, we find the great Rocky Mountains, which run from Alaska southwards for 4800km (3000 miles), reaching into New Mexico. This is an area of rugged natural beauty with snow-dusted peaks, rolling forests, and spectacularly clear,

spring-fed lakes as well as some of the region's most dazzling scenery and wildlife. Sweeping across the mountain tips into the interior, we reach the Great Plains. These vast areas of grassland form the focus of North American rural life and, side by side with more everyday domestic animals, many of the continent's rarest and wildest species can be found here.

Pressing northwards into Canada, we enter the lands dominated by short cool summers and long cold winters. Here, tough Arctic animals must compete, not just with each other, but with the climate and environment to make it through the day. From Florida to Alaska, from Alligators to Polar Bears: the North American journey is one full of danger, conflict — and many surprises.

Alligator

Only found in the furthest reaches
of the southern USA, this giant
predator is one of the region's
biggest reptiles. These territorial
meat-eaters will take a bite out of
almost anything that comes within
reach, and are among the
animal kingdom's best-
documented cannibals.

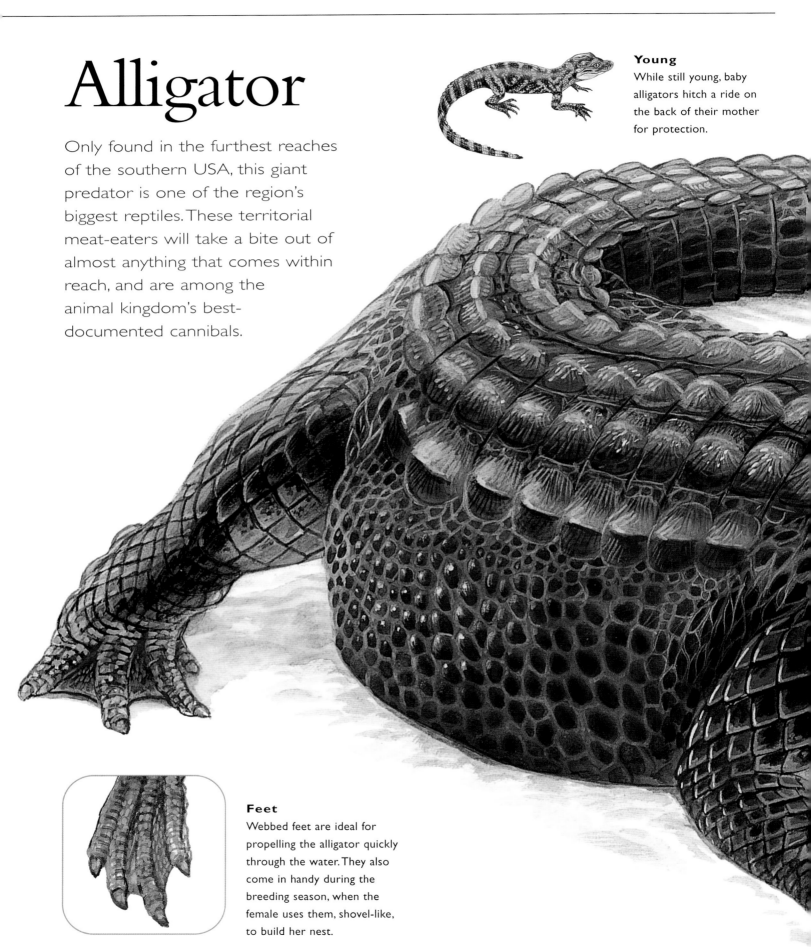

Young
While still young, baby
alligators hitch a ride on
the back of their mother
for protection.

Feet
Webbed feet are ideal for
propelling the alligator quickly
through the water. They also
come in handy during the
breeding season, when the
female uses them, shovel-like,
to build her nest.

Key Facts

	ORDER *Crocodylia* / FAMILY *Alligatoridae* GENUS & SPECIES *Alligator mississippiensis*
Weight	Up to 250kg (551lb)
Length	Male up to 6m (19ft 7in) in the past, now 4m (13ft) considered large; female smaller
Sexual maturity	10–12 years in wild; sooner in captivity
Breeding season	April to May
Number of eggs	35 to 50 per clutch
Incubation period	60–70 days
Breeding interval	1 to 3 years
Typical diet	Birds, small mammals, snakes, fish, turtles
Lifespan	Up to 30 years

Teeth

An alligator's teeth grow continually. Old, worn teeth are eventually replaced when new ones push them out from below.

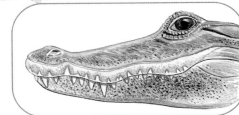

Comparisons

American Crocodiles are found in the southernmost reaches of North America, and down into Ecuador and Columbia. They often share their ranges with alligators and both can sometimes be seen during the hot summer months, basking on the river banks side by side. Compared to the alligator, the American Crocodile has a more wedge-shaped head, and a relatively slender body. This makes it faster in the water. It's no surprise, then, that the crocodile's diet is more aquatic than the alligator.

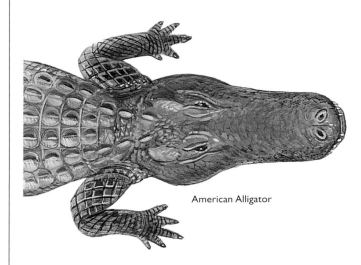

American Alligator

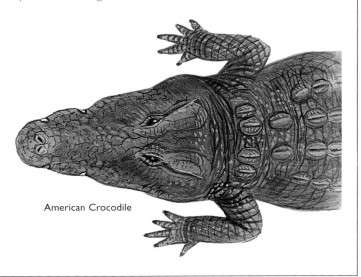

American Crocodile

The American Alligator is much larger than the only other species in the world, the Chinese Alligator. It may not be as big as the American Crocodile, but this latter-day dinosaur has been known to reach 5.7m (19ft).

When is a Crocodile an Alligator?

Since they're closely related to crocodiles, it's not surprising that alligators look very similar. They can, however, be easily identified by their teeth and the shape of their head.

When a crocodile closes its mouth, the fourth tooth in the lower jaw, which is longer than the rest, is visible. In an alligator, this tooth fits into a groove in the upper jaw so that it's hidden when the jaws are closed. The alligator's head is also shorter, with a rounder, blunter snout that makes it look as if it has a permanent, broad-mouthed smile. Yet this congenial appearance is deceptive. In Florida's swamps and everglades, American Alligators sit at the top of the food chain. When young, alligators may eat nothing more substantial than insects and shrimp, but as they grow they're able to take on bigger prey. No one knows for sure how long an American Alligator can live, but some are known to be at least 50 years old, when these grand old 'gators are large enough to tackle cattle, deer and horses. According to the wildlife monitoring tags found in the stomachs of older 'gators, they also regularly top up their protein supply by eating their young. Like all large carnivores, alligators don't need to eat every day. This leaves plenty of time to indulge in their favourite pastime, which is basking on the banks of the river. Alligators seem to enjoy this so much that groups can wear down the river bank and create their own small lagoons by their constant rolling in the mud. Yet cannibalism is such a real danger

Alligator habitats

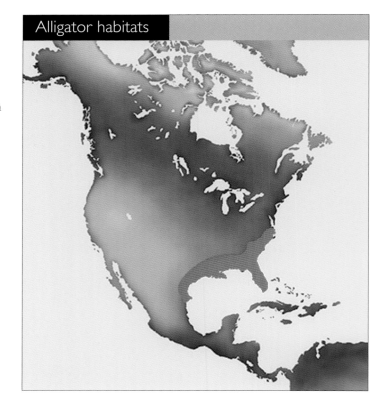

that basking alligators gather in groups with others of the same size, so that no one 'gator can overpower another.

Nest Building

An angry alligator will often roar at an opponent to warn it away, but these fearsome sounding calls are also used for another reason: to attract an mate. Competition to breed during the summer is intense, and males fight each other for dominance. These roars, plus a musky odour, attract interested females. Once the alligators have mated, the female builds a nest, using her jaws to scoop up mud and vegetation. The finished nest is an impressive sight about 90cm (3ft) high. This not only keeps the eggs (20–60) safe, but the rotting vegetation warms them. It takes between two and three months for the eggs to hatch, during which time the female guards them devotedly.

Sun Seekers

There is a well-known story that New York's sewers are infested with alligators. The tale dates back to the 1970s and '80s, when it became fashionable to keep baby 'gators as pets. When the owners grew bored, the story goes, the pets were flushed into the sewers, where they thrived – living on rats and the occasional unfortunate maintenance worker! Such tall tales might raise a smile in New York, but in Florida alligators are no laughing matter. Drainage channels dug to prevent flooding in residential areas have turned into alligator highways, Whether on the streets, in storm drains, on golf courses or in swimming pools, alligators are among the Sunshine State's most regular, unwanted visitors. The bad news is that, while humans might not be their natural prey, even an inquisitive nip from an alligator can be enough to kill.

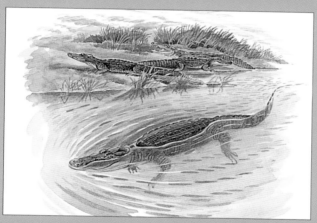

The breeding season begins in March after the alligators have emerged from their winter sleep.

The male roars, both to warn off potential rivals and to attract receptive females.

The male rubs his head against her back, and the two then circle one another in the water.

In the shallow water, the male covers the female with a forelimb and a hindlimb and they mate.

Gila Monster

As the largest lizard in the United States, the Gila Monster has a formidable reputation. Yet much of what we think we know about this seemingly aggressive carnivore is based on myth and superstition. Despite being one of the world's few poisonous lizards, the Gila Monster may not actually be as dangerous as we believe.

Key Facts	ORDER *Squamata* FAMILY *Helodermatidae* GENUS & SPECIES *H. suspectum*	
Weight	4.5–5kg (10–11lb)	
Length	40–55cm (16–21in)	
Sexual maturity	3–4 years	
Mating season	Spring	
Number of Eggs	3 to 13	
Incubation Period	117–130 days	
Breeding Interval	1 Year	
Typical Diet	Young mammals and birds, as well as the eggs of birds and reptiles	
Lifespan	About 20 years	

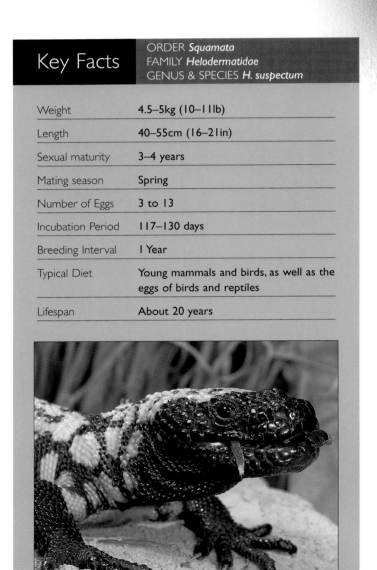

Skin

Life in the dry desert can be tough. To save vital fluids, the Gila Monster's body is covered in beadlike scales that prevent unnecessary moisture loss.

Tongue
Like many members of the lizard family, Gila Monsters use their tongue when hunting – effectively 'tasting' the air for a scent of their prey.

Mouth
With poison-filled ducts, and around 10, sharp, grooved teeth in each jaw, the Gila Monster's mouth is certainly its most effective weapon.

Comparisons

Lizards vary in size and aggressiveness. The Beaded Lizard and the Gila Lizard are only small when compared to the Asiatic Monitor or the gigantic Komodo Dragon, but they are generally considered dangerous because they are the only two species producing poison.

Beaded Lizard

Gila Monster

Asiatic Monitor

Found in Utah and New Mexico and also occasionally in southern Nevada and California, the Gila Monster roams across an extensive range. There are two sub-species: the Banded Gila Monster, which is found more often in the north, and the Reticulated Gila Monster, which prefers the south. However, the monster's distinctive coloured pattern is so variable that clear identification of each sub-species is not always possible.

Tall Tales

When someone appears to be more aggressive than they really are, we often say that 'their bark is worse than their bite', and this is certainly true of the Gila Monster. In the nineteenth century, the belief was that when a gila bit its prey, it wouldn't let go until sunset. Also commonly accepted was the idea that its breath was poisonous. In the *Scientific American* in 1890, one scientist commented that 'the breath is very fetid and…it is supposed that this is one way in which the monster catche …its food…the foul gas overcoming them.' Another in 1907 declared that 'Old settlers here know of many cases of Gila Monster poisoning in which the effect was death.' Like the Komodo Dragon, Gila Monsters do have powerful jaws – and bad breath – but none of these old settlers' tales are true.

In fact, even a gila's poison can be fairly ineffective, as it has no direct means of injecting its venom into its victims.

When a gila bites, venom is expelled from poison glands on either side of its jaw and filtered along grooves in the lizard's teeth. But it's unlikely to be fatal to a human. Nor

Gila Monster habitats

is the venom necessarily fatal to its prey, which may actually succumb to shock.

Fat Tails

A Gila Monster is most at home in a cool, damp environment. Yet it has been able to make its home in the deserts of the southern USA, thanks to some unique adaptations. Most of a Gila Monster's life is spent in underground burrows. As they are diurnal, however, the only way to avoid the heat of the midday sun is to restrict its 'active period' to the cooler months of the year – mostly spring. For about three months, the gilas emerge from their burrows to mate and feed, after which they return underground. Gila Monsters can eat up to a third of their body weight in one meal. Just four of these mega-meals are enough to last them the whole year, and they can eat most of what they need to survive in these months. These food 'reserves' are then stored as fat in the gila's tail, so it's easy to spot a healthy, well-fed monster. It's the one with the tubby tail!

Slow Food

Everything about the Gila Monster's lifestyle is designed to save energy, as the more active it is, the more often it has to leave its cool burrow to feed. This makes them natural foragers rather than hunters. A gila's preferred prey is one that can't run away. High on the menu are baby birds, small reptiles, insects and eggs.

When foraging, gilas may have to cover huge territories to find enough food. Luckily, their active times coincide with periods when there's a ready supply. April and May, for example, is when Gambel's Quails lay their eggs. These birds build nests on the ground and lay up to a dozen eggs at a time. More than enough for a hungry Gila Monster!

The Gila Monster uses its tongue to test for the scent of prey in the area.

Recognizing the smell of a desert cottontail on the wind, the Gila Monster closes in.

The lizard grips its prey with its powerful jaws until the rabbit starts to succumb to the poisonous saliva.

The Gila Monster eats its victim whole after it dies, either from shock or from the poisonous bite.

Grizzly Bear

Key Facts	ORDER *Carnivora* / FAMILY *Ursidae* / GENUS & SPECIES *Ursus arctos horribilis*
Weight	225–320kg (496–705lb)
Length	1.5–2.5m (5ft–8ft)
Shoulder height	90–105cm (35–41in)
Sexual maturity	2–3 years
Mating season	Spring to early summer
Gestation period	About 225 days
Number of young	1 to 4, usually 2
Birth interval	2–3 years
Typical diet	Varied; includes nuts, berries, roots, insects, small vertebrates, fish and carrion
Lifespan	Up to 25

Bears are powerful and able hunters, who can run at 56km/h (35mph) and bring down a large mammal, even a human, with a single blow. They're not natural man-eaters, but will attack with ferocity if threatened.

Paws

Bears, like humans, are one of the few animals who walk on the soles of their feet, rather than just the toes.

Teeth

Grizzly Bears are omnivorous – they eat animals as well as vegetables – so they are equipped with teeth that can both grind and tear.

Comparisons

Despite being rare in Europe, bears are still relatively common in the far reaches of North America. In general, the size of bears tends to increase as the species travels north. So, the American Black Bear, found in wooded areas of North America, is amongst one of the continent's smallest, and the Polar Bear, found in Alaska and Siberia, is the largest.

American Black Bear   Grizzly Bear   Polar Bear

Grizzlies get their name not from their temperament, but their colour, as their fur is generally brown, tipped with white, making them look 'grizzled' (streaked with grey). Hugely powerful, grizzlies can be distinguished from other bears by their size, a hump between their shoulders, and their long curved claws.

All-Round Athlete

Like most members of the family *Ursidae*, Grizzly Bears are omnivorous. They get much of their energy from roots, bulbs and tubers, which they dig up using their long claws. However, as they need to feed to survive, breed and hibernate, a grizzly is unlikely to turn down any readily available meal, which includes fresh meat, carrion and even garbage. All bears have an incredible sense of smell, and food left lying around will quickly attract their attention – so much so that campers are advised to keep all 'smellables' well away from sleeping areas to prevent unwanted encounters with scavenging bears.

Long, cold winter months, when food is scarce, present the biggest survival challenge to most animals. The solution for many is hibernation – a sleeplike state during which the animal lowers its body temperature and heart rate to save energy. Zoologists are divided as to whether bears go into full hibernation or simply become 'dormant' during the winter. Whatever the answer, this period of prolonged sleep is one for which for a grizzly must be well prepared – and it's here that its natural omnivorous nature comes to the fore.

Countdown to Hibernation

During 'hibernation', Grizzly Bears can burn up a million calories. By the spring, when they are ready to leave their dens, they may have lost a third of their body weight, so it's important to start piling on the calories immediately. Luckily, grizzlies are well equipped to take advantage of a range of seasonal goodies. In April, roots and grasses make up the bulk of the menu. By May, fresh meat is available and grizzlies go on the hunt, tracking down prey such as young deer with surprising speed and agility. Summer brings a salmon bonanza as the fish swim up stream to spawn (lay eggs). Grizzlies are expert fishermen and may

Grizzly Bear habitats

catch a dozen salmon a day, which adds valuable protein to their diet. By the autumn, grizzlies turn to berries and nuts for a final calorie boost. When winter comes, this varied and flexible diner will be ready once again to sleep its way through the long, dark nights.

At Risk

In the past, Native Americans used to hunt Grizzly Bears for food, either digging pits to trap them, or lassoing them from horseback, as described by Nelson Lee in his memorable story of wilderness life, *Three Years among the Comanches* (1859):

'A part of the Indian's equipage is the lasso…and in the skill…with which they throw it, they far excel the Mexicans. On this occasion…a couple of them galloped out, one dextrously throwing the noose over Bruin's neck and twitching him onto his back, the other… throwing another over his hind legs, thus subjecting him to a most uncomfortable stretch as they pulled in opposite directions.'

Once, grizzlies roamed throughout the Americas, from the Arctic Circle to central Mexico. As settlers moved into grizzly territory, however, they were increasingly killed, not for food but simply because they were considered a nuisance. They are now extinct in much of their former range, apart from national parks such as Yellowstone, 'recovery areas' like Bitterroot, and in the far north of Alaska, British Columbia and the Yukon. Nevertheless, they're still regularly killed by trophy hunters. Grizzlies breed particularly slowly, and may not have cubs at all if food is scarce, so there's a real worry that, very soon, grizzlies may die out entirely.

In spring, bears are drawn to the rivers as the salmon attempt to swim upstream, waiting for the fish to jump.

Some bears wait with their head near the water and snap at the salmon when they jump.

Others prefer to launch their whole body at a fish before it has a chance to leap, catching it off guard with a belly flop.

The flesh is rich in protein and the skin has a high fat content, perfect to replenish stores after 'hiberation'.

Puma

Pumas are masters of the ambush. Exploding from a crouch, an adult uses its powerful hind legs to pounce on prey, jumping 4.5m (15 ft.) from a standing start, and over 13.7m (45 ft.) when running. Such athleticism is rarely wasted: 8 out of 10 hunts end in a kill, making the puma a deadly predator.

Eyes
Generally, prey species have eyes at the side of their heads, which gives them a wide field of vision. As hunters, pumas have eyes in front – allowing them to better focus on prey.

Mouth
Ridges on the roof of the puma's mouth help it to grip onto prey.

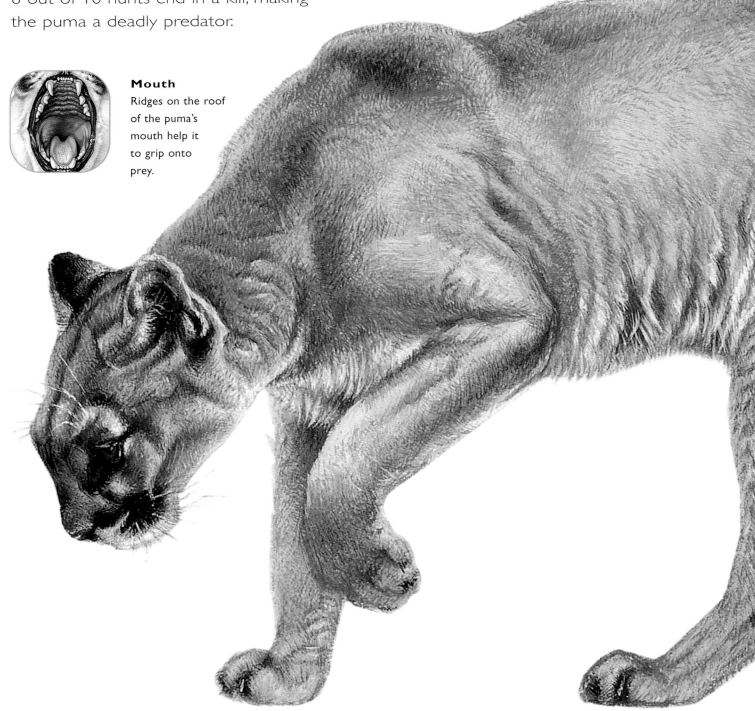

Legs

With large, powerful leg muscles and an ideal size-to-weight ratio, pumas make fast and efficient predators.

Key Facts

ORDER *Carnivora* / FAMILY *Felidae* / GENUS & SPECIES *Felis concolor*

Weight	Male 65–105 kg (143–231lb). Female 35–60 kg (77–132lb).
Length Head and Body	Male 1–1.9m (3ft 3in–6ft 2in). Female 0.9–1.5m (3–5ft).
Tail	50-80cm (20–31in)
Shoulder	60-70cm (24–28in)
Sexual maturity	2-3 years
Mating season	Varies with habitat
Gestation period	90-95 days
Number of young	1 to 4
Typical diet	Mammals, birds and insects
Lifespan	20 years

Comparisons

Looking like an overgrown, stocky domestic cat, the Pampas Cat – like the puma – can be found in a wide range of habitats, from the Equator to Argentina. It's also a nocturnal hunter, but, due to its size, prefers small prey like rodents and birds.

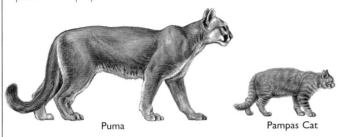

Puma   Pampas Cat

When European settlers first saw the puma, there seems to have been quite a bit of confusion about what species it belonged to. When they're born, puma cubs have spots, like a leopard. These vanish at around 6 months, leaving the puma with an almost monochrome (one colour) coat, which is why it's classified scientifically as *Felis concolor*, meaning cat of one colour.

The Name Game

Despite its scientific classification, the puma has coat that varies in colour depending on which of the 30 or so sub-species it belongs to. In fact, pumas may be any colour from silver grey to reddish-brown. Coat-length varies too, being shorter in warmer regions, and longer in the cooler North. This is one of the reasons, perhaps, why pumas have been know by many names, including brown tigers, purple feather, silver lions and, in the east of the United States, panthers, which is a name applied to numerous members of the cat family, especially all black varieties.

Pumas are widespread throughout North, South and Central America, but tend to favour mountain areas, which is how this large cat got some of its other names, like catamount ('cat of the mountain') and cougar, meaning mountain lion, as some people believed that pumas were really female African Lions. Some of the puma's bewildering array of names clearly reflect its physical traits. During the breeding season, female pumas let out a blood-chilling scream to attract a mate. Hence the name 'mountain screamer'. Other names play on their reputation and prowess as hunters, such as deer tiger, sneak cat and mountain demon. Today, cougar, mountain lion and puma are all accepted names.

Messy Messages

Pumas are solitary hunters and, depending on conditions, each male may cover a territory as large as 259 square kilometres (100 sq. mi.), which will usually overlap the territories of several females. This makes encounters with other males rare, but most pumas will still go to great lengths to mark out the boundaries of their range. Many male animals do this. Birds use song to declare their 'ownership' of a specific territory. Other animals have musk glands, which they rub against trees and rocks to let competitors know that they've passed this way. Pumas scent mark, 'spraying' urine along the edges of their territory. They will also sometimes pile up leaves and dirt, which they 'tag' with faeces. This may sound odd, but marking the boarders of their territory, makes good sense. It keeps out intruders and reduces unwanted conflict.

Unlike its prey, the puma comes to know its territory intimately, giving it an advantage during a hunt.

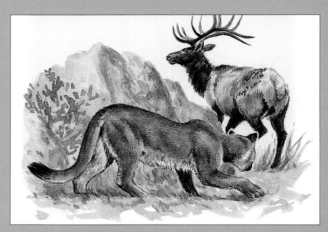

The puma carefully moves in on the elk, often freezing for some time if the elk looks around.

Puma habitats around the world

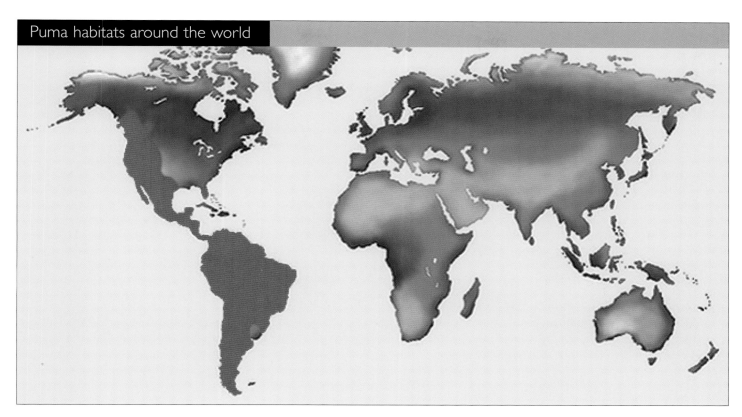

Predator and Rival

Like all members of the Family *Felidae*, pumas are great athletes, and much of their skill and gymnastic ability is dedicated to the hunt. Pumas will eat a wide range of animals, relying on hares, porcupines and rats for a large part of their calorie intake. Their most regular prey are deer and big-horn sheep, which tend to be slightly larger on average than a fully grown male puma. And if the opportunity arises, they'll also tackle elk, moose, horses and domestic cattle. Recently, pumas in parts of California have been responsible for an increasing number of human deaths. Numbers of this great cat have been rising steadily since a hunting ban was introduced in 1972. There are still only around 6000 pumas in the state, but the human population has increased by 25% (to 30 million) in the last 10 years. Once again, humans are encroaching into areas that were once the preserve of wild animals, and deaths are the inevitable consequence.

When the puma is close enough, it makes a sudden explosive leap towards the elk.

If the initial attack does not snap the elk's neck, the puma closes its jaws around the neck and strangles its prey.

Rattlesnake

The Western Diamond Rattlesnake has been a symbol of America since prehistoric times. Equally at home in water, desert or on grassy plains, this large reptile is perfectly adapted for a life of hunting and killing. Even small examples of this species are armed and dangerous, since they are born complete with fangs and venom.

Female tail
The rattle is this snake's most prominent feature. In females, this well-known warning device tends to be shorter than in males.

Fangs

In common with other members of the viper family, rattlesnakes have long fangs set on 'elastic-like' hinges. These spring forward when the snake opens its mouth, and fold back when its closed.

Key Facts

ORDER *Squamata* / FAMILY *Crotalidae*
GENUS & SPECIES *Crotalus atrox*

Weight	Up to 6.8kg (15lb)
Length	Up to 2.1m (6ft 9in)
Sexual maturity	About 3 years
Mating season	Spring
Incubation period	About 165 days
Number of young	Up to 46, but usually 10 to 20
Birth interval	2 years
Typical diet	Small mammals, such as mice and voles; also ground squirrels, rabbits and prairie dogs, small birds, frogs and lizards
Lifespan	Up to 24 years in captivity, usually much less in the wild

Comparisons

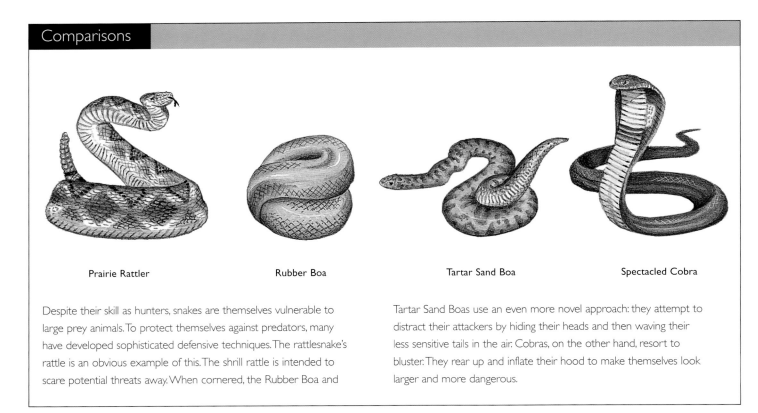

Prairie Rattler

Rubber Boa

Tartar Sand Boa

Spectacled Cobra

Despite their skill as hunters, snakes are themselves vulnerable to large prey animals. To protect themselves against predators, many have developed sophisticated defensive techniques. The rattlesnake's rattle is an obvious example of this. The shrill rattle is intended to scare potential threats away. When cornered, the Rubber Boa and

Tartar Sand Boas use an even more novel approach: they attempt to distract their attackers by hiding their heads and then waving their less sensitive tails in the air. Cobras, on the other hand, resort to bluster. They rear up and inflate their hood to make themselves look larger and more dangerous.

Like other rattlesnakes, the Western Diamond makes its home among small communities of rabbits, gophers or prairie dogs, which comprise the bulk of its diet. Yet this skilled and stealthy predator will travel vast distances for a meal, ambushing and poisoning its victims before then swallowing them whole.

Too Big for their Boots

The Western Diamond Rattlesnake typically grows to between 60cm and 1.5m (2–5ft) in length. Like all snakes, however, they continue to grow throughout their life. No one knows for certain how long a rattlesnake may live, but older snakes, between 15 and 20 years old, may be as long as 2.1m (7ft). The reason that snakes are able to continue growing is that they shed their skin regularly.

A snake's skin is made up of thousands of dry scales. The outer layer of scales is actually dead, but beneath this is a 'live' layer, where new scales are made to replace those that become worn out. How often a snake 'moults' depends on its age and how active it is. Young snakes tend to grow quickly in the first few years, and a Western Diamond Rattlesnake may moult two or three times a year during its early life. Every time it sheds its skin, a new segment is added to the rattle. It used to be believed that, just like counting tree rings, the number of segments in a snake's rattle was an indication of its age. However, the rate of shedding varies, and rattles do get damaged, so this method is no longer considered useful.

Keep Away!

All rattlesnakes are born with a bonelike tip on the end of their tail called a rattle, which is made up of hardened keratin (the same substance that forms hair and nails). Rattles are purely defensive. Rattlesnakes may be highly

Western Diamond Rattlesnake habitats

poisonous, but thousands die every year because larger animals stand on them. So, when rattlesnakes are alarmed, they'll hiss loudly and shake their tails to produce a 'rattle' that warns others to stay well away. Although rattlesnakes bite more people in North America than any other snake, they will often follow this warning with a 'dry', venomless bite that is designed to alarm, rather than poison. Old snakes in captivity may have rattles with up to 25 segments, but the bigger the rattle gets, the more the sound is deadened, and the less effective it becomes. The optimum number of rattles is eight. This will produce a warning sound that can be heard up to 1m (3ft 3in) away.

Added Extras

When a snake flicks out its forked tongue, it isn't trying to cool down, like a dog: but is actually smelling the air. The snake's tongue is a highly sensitive instrument that is able to pick up minute particles in the air. These are transferred to specialized pits called Jacobsen's Organs on the roof of its mouth. Jacobsen's Organs translate chemicals from the air into sensory information that the snake can then use to track down prey. This is especially useful for finding dying prey that they have previously poisoned.

The Western Diamond Rattlesnake, however, in common with many species of pit vipers, has an additional sensory organ that turns them into deadly accurate hunters. These organs are located in dips on either side of the head. By moving its head from side to side, the pit organs can detect the smallest change in air temperature. A 'hot' spot invariably means that a warm body is close by. Pit organs are so effective that rattlesnakes can even find their prey in the dark, especially in burrows.

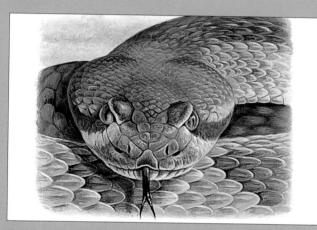

A skilled nocturnal hunter, the snake uses a variety of means to locate its prey, including infrared signature and chemical scent as well as sight.

The snake can detect the body heat of prey in the darkness up to 50cm (20in) away, lending it a distinct advantage over its victim at night.

With a striking distance up to half its total body length, the prey has no chance to turn and flee.

A lethal dose of poison is administered through the long fangs sunk into the victim.

Wolverine

Wolverines are sometimes known as 'gluttons' and this nickname is well deserved. Given the opportunity, a wolverine will attack almost any animal. In an amazing display of ferocity and power, a male wolverine weighing between just 11 and 18kg (24–40lb) will prey on a fully grown caribou weighing 113–320kg (249–705lb).

Jaws

Large, muscular jaws give the wolverine an incredibly powerful bite. They can crunch through frozen carrion with ease.

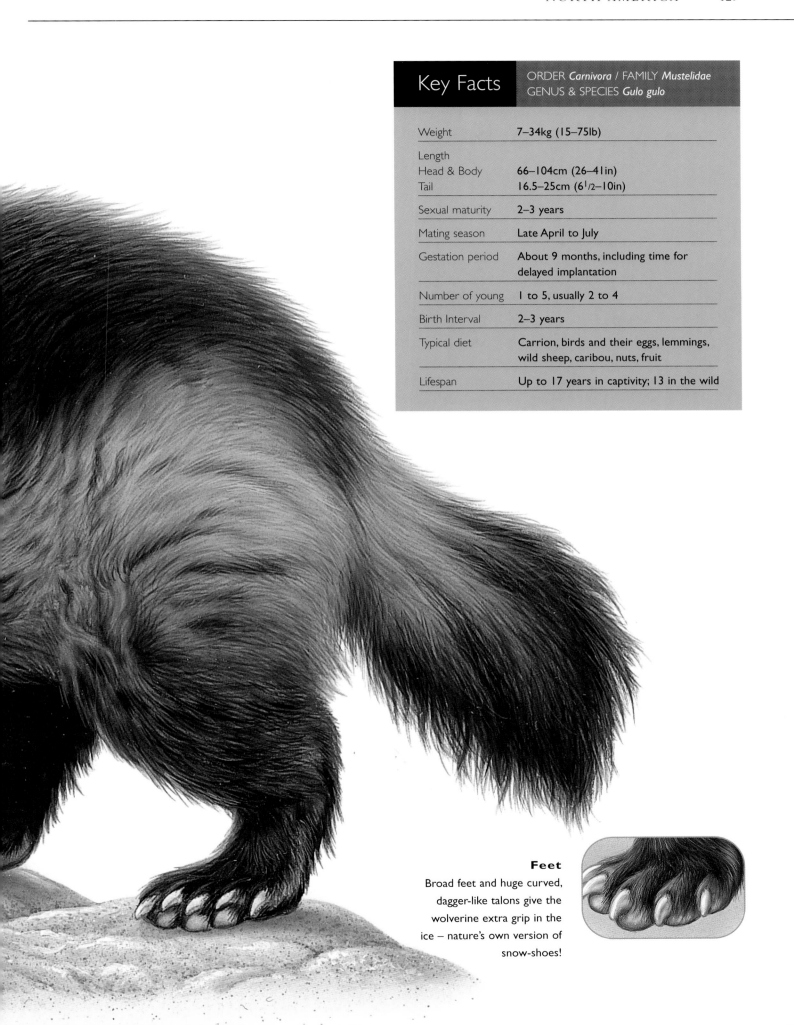

Key Facts

ORDER *Carnivora* / FAMILY *Mustelidae*
GENUS & SPECIES *Gulo gulo*

Weight	7–34kg (15–75lb)
Length Head & Body Tail	66–104cm (26–41in) 16.5–25cm (6½–10in)
Sexual maturity	2–3 years
Mating season	Late April to July
Gestation period	About 9 months, including time for delayed implantation
Number of young	1 to 5, usually 2 to 4
Birth Interval	2–3 years
Typical diet	Carrion, birds and their eggs, lemmings, wild sheep, caribou, nuts, fruit
Lifespan	Up to 17 years in captivity; 13 in the wild

Feet
Broad feet and huge curved, dagger-like talons give the wolverine extra grip in the ice – nature's own version of snow-shoes!

Wolverine habitats

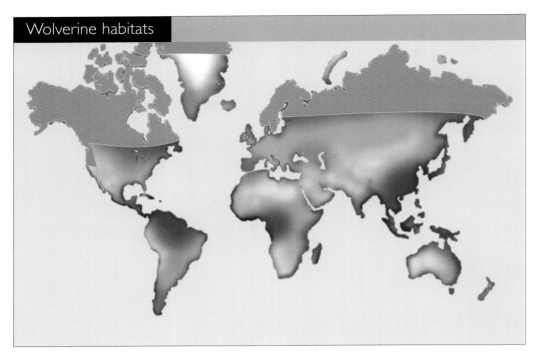

effective predators. Wolverines are fierce and fearless. Although they enjoy a varied diet of small mammals, birds, carrion and, occasionally, plants and berries, they regularly prey on animals as much as ten times their size. Anyone familiar with Marvel Comics' Wolverine character will know that he has massive claws. Real wolverines are similarly well endowed, and use their huge curved, dagger-like talons to great effect when hunting. Jumping on a caribou's back, the wolverine will then dig in his claws and begin to slice and render the animal's flesh. Once the deer falls to the ground, overcome by shock and blood loss, the wolverine will literally tear the carcass apart, burying large hunks in the snow to eat later.

Native North Americans revered the wolverine. In their stories, he was referred to as the great 'trickster-hero' of the spirit world, but in the real world the wolverine is just as remarkable.

Marvel-ous!

Wildlife illustrator and naturalist *Ernest Seton* (1860–1946) gave this description of the wolverine: 'Picture a weasel – and most of us can do that, for we have met that little demon of destruction, that small atom of insensate courage, that symbol of slaughter… picture that scrap of demonic fury, multiply that mite some fifty times, and you have the likeness of a wolverine.'

While Seton clearly wasn't a fan of wolverines, he was nevertheless fairly accurate in his assessment of the qualities that make wolverines such dangerous and

Snow Problem

The wolverine is the largest member of the family Mustelidae, which includes weasels, otters, badgers and skunks. Powerfully built, with an almost bear-cub appearance, the wolverine has made its home in the far, snow-bound north – specifically the Yukon, Alaska and the Northwest Territories, as well as some parts of northern Europe.

Fossil records show that the wolverine used to be much more widespread and it is believed that their ancestors were 'pushed' into the Arctic extremes when human settlers moved into their habitats. However, they've

Comparisons

Weasels, otters, badgers, skunks and wolverines all belong to the Family *Mustelidae*, so called because of the strong 'musky' odour that these species give off. Apart from this unfortunate trait, members of this family share many other features. They all, for example, tend to be fast and active predators. They also generally have slender bodies, short legs and small heads. The smallest of this family is the Least Weasel, while the wolverine and the Sea Otter rank amongst the largest.

Sea Otter   Wolverine   American Marten   Least Weasel

adapted wonderfully to the demands of an Arctic existence. Their coats, for example, are thick and glossy, which is perfect for heat retention. Their legs are short and they walk with a 'plantigrade' gape. Most animals walk on their toes but, like humans and bears, wolverines put their entire foot on the ground, which gives much better traction, especially on uneven or icy surfaces. Their skull and jaws too are especially suited to icy living. In fact, they are so robust that a wolverine can crunch through frozen carrion with ease.

Birth and Death

Typically a wolverine's life is a fairly solitary one, except during the breeding season, when males tend to stay close to the females. Like Polar Bears, female wolverines can take advantage of delayed implantation to ensure that their litter of two or three young ('kits') are born at the most opportune time. This is usually March, when food is plentiful, and ensures that the young wolverines will be weaned and approaching adulthood by the time they have to face the challenge of their first winter. The kits are born with white fur, which slowly changes to dark brown over the course of the year. It's common for animals' reproductive cycles to be timed to coincide with periods of abundant food. Even so, many kits will still die of starvation before they're weaned.

Adult wolverines are so ferocious they have few natural enemies – generally only pumas and bears will risk taking them on. However humans also present a real threat to the species: wolverines were often killed for their fur in the past, and even today they are still in danger from trophy hunters and farmers.

The wolverine searches for prey over a wide area, using its acute sense of smell as a guide and spots a caribou.

In the deep snow the wolverine's broad paws act like snowshoes, giving it a distinct advantage over many prey.

Unable to flee through the deep snow, the caribou is helpless as the wolverine leaps upon it, biting the neck as it holds on.

Exhausted through exertion and blood loss, the caribou dies. After the wolverine has eaten, it will dismember the remains of the carcass and bury it for later.

Polar Bear

Polar Bears are the largest land carnivore in the world. Fast, powerful and agile, a Polar Bear can run at over 56km/h (35mph), swim at 10km/h (6mph), and jump an amazing 2.13m (7ft) straight out of water in search of a meal.

Polar bears fighting
In the frozen North, resources can be scarce, making fights between bears particularly vicious and bloody affairs.

Key Facts

	ORDER *Carnivora* / FAMILY *Ursidae* GENUS & SPECIES *Ursus maritimus*
Weight	Male up to 725kg (1598lb); female up to 249kg (549lb)
Length	2.4–3m (7ft 10in–9ft 10in)
Sexual maturity	Usually 4–5 years. Most males do not mate until 8–10 years old
Mating season	Mainly April to May
Gestation period	195–265 days, which includes a period of delayed implantation
Number of young	1 to 4, usually 2
Birth interval	3–4 years
Typical diet	Mainly seals, especially the ringed seal
Lifespan	Up to 30 years, but usually 15–18

Teeth
Polar Bears are well equipped for an almost exclusively meat diet. Their 42 teeth include 4 large canines for piercing flesh.

Paws
A Polar Bear's huge paws help spread its weight on the ice, a little like wearing snow-shoes.

Polar Bears are the most carnivorous members of the bear family. As their long, sharp teeth testify, they live almost exclusively on a diet of meat. These gigantic mammals are among the Arctic's top predators, although attacks on humans are, thankfully, rare.

Hot Stuff!

Life in the far northern hemisphere can be challenging, but the Polar Bear is superbly adapted for survival in the cold, ice-bound extremes of Russia, Greenland, Canada and Alaska, where it makes its home.

In these regions, temperatures can fall rapidly, but a Polar Bear's body is so well insulated that when it runs it's actually in danger of over-heating. The reason for this is the bear's remarkable body design. Outermost is a double mantle of fur: an undercoat of fine white hair and an outer coat comprised of long 'guard' (protective) hairs, which are hollow to provide additional buoyancy in water. Beneath the fur is the Polar Bear's skin, which is black, since black is more efficient at absorbing heat than white. Under these protective layers is a final insulating layer of fat, which can be as thick as 10cm (almost 4in). Indeed, the Polar Bear's body is so good at retaining heat that, if we were to film it using a heat-sensitive camera, it would be almost invisible.

Perfectly Adapted

The Polar Bear's scientific name is *Ursus maritimus*, meaning 'sea bear', and they are in fact excellent swimmers. Polar Bears spend most of their lives in Arctic coastal areas, where they can often be seen hitching lifts on passing ice flows. On the ice, Polar Bears aren't as agile or

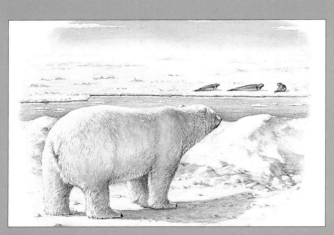

The Polar Bear spots a fine meal: a seal hauled out on the ice. The bear evaluates the terrain to find the best route of attack.

The bear slowly swims up an open channel in the ice, keeping a low profile as it closes the distance.

Caught unawares, the seal has no chance of escape as the Polar Bear leaps onto the ice.

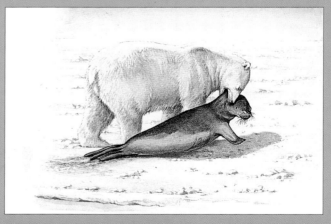

The Polar Bear drags the seal away to eat the blubber and flesh. One seal must be consumed every six days for the Polar Bear to maintain its body weight.

fast as Brown Bears, but they make up for this in the water. Using their powerful webbed forepaws to propel themselves forwards, a Polar Bear can swim up to 96km (60 miles) without a rest.

Living in such extreme conditions requires a lot of fuel to keep warm, even for an animal with a body as efficient as the Polar Bear's. To survive, an average-sized bear must eat 2kg (4.4lb) of fat every day. An adult seal can provide a Polar Bear with enough calories for a week, which is why they rank among this great carnivore's favourite food. Being excellent swimmers, Polar Bears have no problems hunting in the water, but they have learnt that the best way to catch a meal is with a little patience. Seals make dozens of holes in the ice, which they use to come up for air. A Polar Bear's sense of smell is seven times more acute than a bloodhound's, allowing them to smell a seal through 1m (3ft 3in) of ice. So all a hungry Polar Bear has to do is to scent out any seals in the area and wait by a breathing hole for them to surface.

Polar Bear habitats

Practical Parenting

Of all the Polar Bear's adaptations, perhaps the most amazing occurs during mating. Polar Bears mate between March and July, but it may not be until September that the female becomes 'officially' pregnant. A process called delayed implantation means that it can be up to six months before the fertilized ovum (egg cell) attaches itself to the wall of the uterus, where the baby will develop.

Unlike male Polar Bears, pregnant females spend long periods during the winter asleep. If the female doesn't gain enough weight before this happens, the embryo will not attach to the uterus. This remarkable form of 'birth control' ensures that female Polar Bears become pregnant only when there is enough food to support herself and her cubs.

Comparisons

Sometimes being big can be a disadvantage. While Polar Bears need an immense amount of fuel to keep their huge bodies warm, Arctic Foxes save energy by being small. Both animals, however, appreciate the advantage of proper insulation. Like the Polar Bear, the fox is covered in a mass of thick fur, which keeps out the cold.

Arctic Fox

Polar Bear

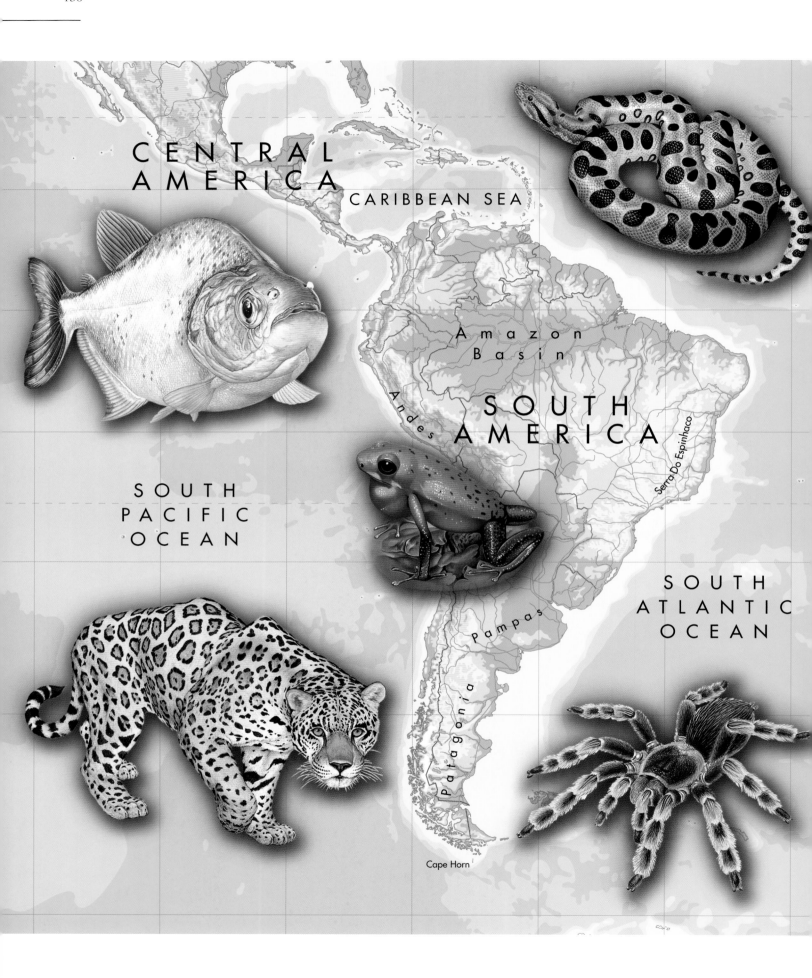

CENTRAL
AMERICA

CARIBBEAN SEA

Amazon
Basin

Andes

SOUTH
AMERICA

Serra Do Espinhaço

SOUTH
PACIFIC
OCEAN

SOUTH
ATLANTIC
OCEAN

Pampas

Patagonia

Cape Horn

South and Central America

This is a region of truly record-breaking excess. In this vast, tear-shaped land mass, which divides the Pacific and Atlantic Oceans, it is possible to discover deserts where rain hasn't fallen for decades, rainforests where the tree canopies are so thick that daylight never reaches the forest floor, and high snow-laden mountains where few but the most nimble and rugged animals ever venture.

If we were to take a whistle-stop jaunt along the coast of what is known collectively as Latin America, our tour of this amazing region would begin in Mexico. This high, thin plateau-nation starts at the borders of the USA and gently curves southwards for 2000km (1250 miles) until it meets its neighbours, Guatemala and Belize. Sweeping on through Nicaragua, Costa Rica, El Salvador and Panama – which form the core of the Central American landmass – our tour would reach South America. This is home to a quarter of all the world's animals and is a land of unbelievable abundance. It's here that we would find the world's longest mountain chain, the Andes, and the world's largest tropical rainforest, the Amazon. Covering around 6 million square kilometres (2.3 million square miles), the Amazon is home to more species of animals than can be found in the rest of South America together. This includes some of nature's most remarkable giants, such as the heaviest snake, the anaconda, plus thousands of species that can't be found anywhere else. From flesh-eating fish to poisonous frogs, this incredible land is home to some of the rarest, most surprising and dangerous of all animals.

Green Anaconda

Perfectly adapted for a life amongst the swamps and streams, this gigantic snake uses stealth, cunning and sheer raw power to make a meal of carnivores and herbivores alike. If lions are kings of the jungle, then Green Anacondas are truly the monarchs of the rainforest.

Eyes
As they don't have lids, a snake's eyes are protected from damage by a thin membrane called a brille.

Pelvic Claws
Proof that snakes were once lizards can be found at the tip of an anaconda's tail. Here are tiny clawlike barbs, which are actually all that remains of their legs.

Key Facts

ORDER *Squamata* / FAMILY *Boidae* / GENUS & SPECIES *Eunectes murinus*

Weight	Up to about 135kg (298lb)
Length	5.4–7.6m (17ft 8in–25ft)
Sexual maturity	4 years
Mating season	Chiefly December and January
Gestation period	6–7 months
Number of young	Varies with the size of the adult
Birth Interval	1–2 years
Typical diet	Aquatic and semi-aquatic animals, including fish, caimans, capybaras and waterbirds
Lifespan	Up to 20 years in captivity

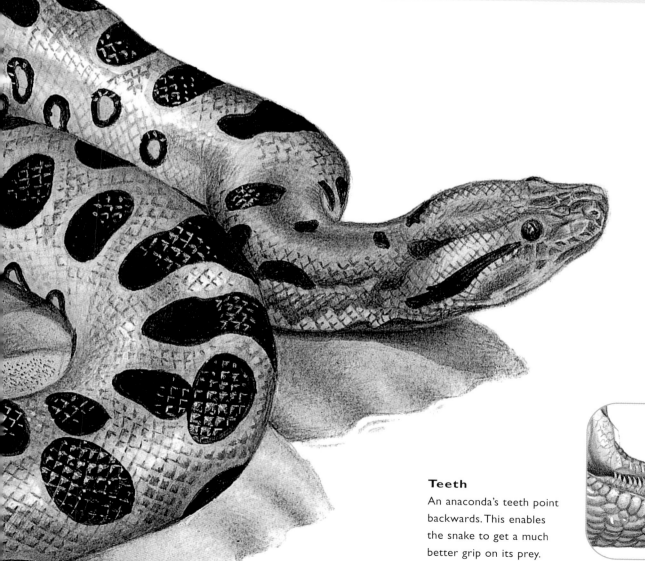

Teeth

An anaconda's teeth point backwards. This enables the snake to get a much better grip on its prey.

Comparisons

In many ways, the Green Anaconda can be seen as the New World equivalent of the Old World Reticulated Python. Both of these snakes grow to gigantic lengths. Both are natural jungle dwellers, who are equally at home in the tree tops or water. More importantly, both are constrictors, which means that they kill their prey by slow suffocation. It used to be believed that constrictors actually crushed their prey to death, because some species 'spit out' their prey's fur once they have digested the meal.

Green Anaconda

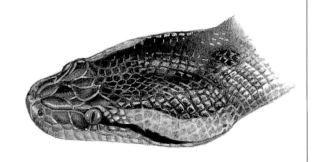

Reticulated Python

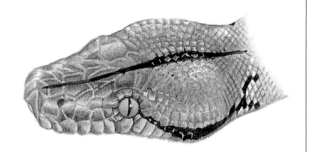

Found throughout South America, especially in the great Amazon basin, Green Anacondas are at home anywhere there is water, from forest to grasslands.

Big Snake, Tall Tales

The name anaconda comes from the Tamil word for 'elephant killer'. Such a name is not surprising for an animal that has been the subject of so much myth and speculation over the centuries. Despite regularly reaching lengths of 5 to 8m (16–27ft), the anaconda isn't the longest snake in the world. That record is still held by the Asian Reticulated Python. It is, however, the biggest snake in terms of weight, often measuring more than 30.4cm (12in) in diameter. Yet tales of monstrous anacondas persist. South American tribespeople, for example, claim that 24m (80ft) examples are a regular sight.

One of the most spectacular tales of these super-sized snakes was told by the English explorer Colonel Percy Harrison Fawcett (1867–1925). Fawcett was the man whose work inspired Sir Arthur Conan Doyle (1859–1930) to write the story of *The Lost World*, in which a group of travellers discovers a land where dinosaurs still exist. In one thrilling account of his travels, the colonel claimed to have killed an anaconda 18.8m (62ft) long. Fawcett disappeared in 1925 during a trip to Bolivia – perhaps a victim of one of these giants?

Water Boas

Anacondas are sometimes called water boas, because they spend most of the time in or close to water. Like crocodiles, they're well adapted to this semi-aquatic life. They have eyes and nostrils on the very top of their head so they can lie almost totally submerged yet still breathe. While slow and sluggish on land, they're fast swimmers and can hold their breath for up to 10 minutes. Not surprisingly then, fish and caimans (a relative of the alligator) comprise a large part of their diet.

Green Anaconda habitat

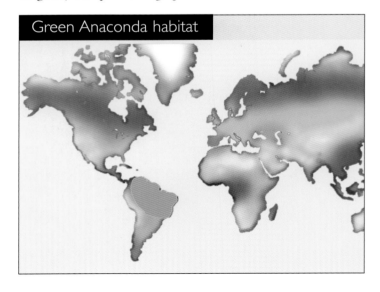

Unaware of the danger nearby, a young caiman comes to the waters edge. The anaconda moves in closer.

The anaconda strikes, latching on to its startled prey with its backward-facing teeth.

The caiman is quickly enveloped by the long muscular body of the anaconda, which then squeezes it to death.

The snake must dislocate its jaws to consume such a huge meal, gradually forcing it down.

Anaconda catch their prey by ambush. Using their natural camouflage, they lie in wait in shallow waters and grab any passing animal. Since they're constrictors, they sink their teeth into their victim, wrap their coils around its body and slowly either suffocate or drown it. This includes any animals that come close to the water to drink. So, deer, jaguars and capybara are also regular additions to the menu. There are scant reliable accounts of anacondas eating humans, mainly because few people live close to their natural rainforest homes, but herpetologists have reported being attacked by hungry anacondas. The fact that these great snakes might not be as big as we think, doesn't make them any less dangerous. They're still one of the most powerful snakes in the world and are opportunistic hunters, who'll tackle any animal they can swallow.

Mothers and Fathers

Anacondas are naturally solitary, coming together to mate during April and May. Attracted by a chemical given off by the female anaconda, up to 12 males may form a 'breeding ball' with the female. This mass of writhing snake flesh is a remarkable sight. For between two and four weeks, the males engage in a continuous wrestling match for the opportunity to mate with the female. Once this mammoth mating session is over, the males return to their bachelor lifestyles, leaving the female to literally carry the babies. At this time, a pregnant female is quite vulnerable to predators, as the additional weight of the embryos slows her down. After six months, the female will give birth to up to 40 live young. These amazing babies can be 60cm (24in) long at birth and are ready within minutes to swim, hunt and feed.

Jaguar

While most big cats kill their prey with a bite to the throat, a jaguar's jaws are so strong that it delivers the finishing bite to the head, piercing the skull of its victim to bring almost instantaneous death. In fact, the native name for this powerful cat is 'yaguara', meaning 'the beast that kills with one bound'.

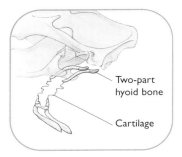

Hyoid bone
Vibrations along the hyoid bone make the familiar cat's roar. Jaguar's grunt rather than roar, but no one knows why.

Key Facts	ORDER *Carnivora* / FAMILY *Felidae* GENUS & SPECIES *Panthera onca*
Weight	Male 90–120kg (198–265lb); female 60–90kg (132–198lb)
Length Head & body Tail	1.1–1.8m (3ft 7in–5ft 11in) 45–75cm (27–30in)
Sexual maturity	2–4 years
Mating season	All year in the tropics; seasonal elsewhere
Gestation period	93–107 days
Number of young	1 to 4; usually 2
Birth interval	About 2 years
Typical diet	Wide variety of mammals; also turtles, fish, caimans and farm livestock
Lifespan	Up to 22 years in captivity

Claws
Long, curved, retractable claws are used
to pin and hold prey while the
jaguar's massively strong jaws
deliver the killing blow.

Comparisons

Looking rather like a miniature leopard, the margay is one of Latin America's smaller wild cats. Although it shares much of its habitat with the larger, predatory jaguar, these 2 proficient killers are able to live side by side with little conflict. The reason is that the margay has a semi-arborial life-style – it spends much of its life amongst the trees. In fact, the Tree Ocelot (as it's also known) is such a skilled climber that much of its diet is made up of birds and squirrels.

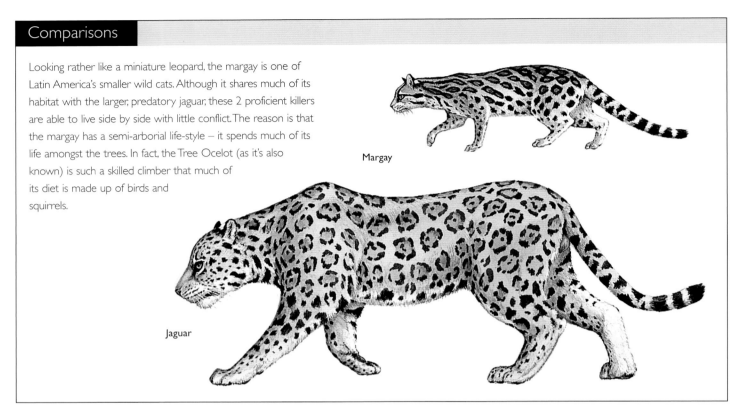

Margay

Jaguar

In ancient Mayan mythology, gods took the form of animals. This included the sun god, who was transformed into the form of a jaguar during his visits to the earth. The Maya considered the jaguar's beauty, grace and strength so dazzling that it simply had to be heaven-sent!

Jaguar habitats

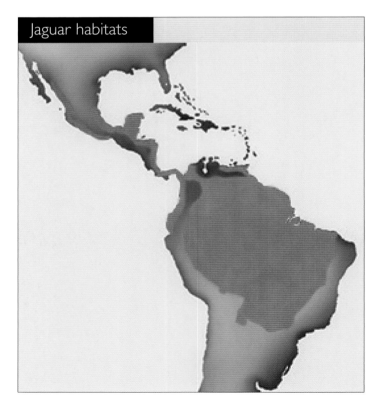

New World, Old Design

The jaguar is the New World equivalent of the Old World leopard. While similar in appearance, its size distinguishes this great dappled cat from its Asian cousin. The jaguar is the third largest of all cats, and the biggest in the western hemisphere. Weighing twice as much as a leopard, with a large head and stocky forelegs, this beautiful and robust animal has few natural enemies apart from man. During the 1960s and '70s, about 18,000 jaguars a year were killed for their fur, and numbers still haven't really recovered. Once common throughout the Americas, including California and New Mexico, jaguars are now primarily found in Brazil, Paraguay and Belize, where areas such as the Cockscomb Basin Wildlife Sanctuary provide a protected environment for these spectacular and skilled hunters. There are currently believed to be as many as eight sub-species of jaguar, although they cover such a huge range that it is difficult for zoologists to be sure.

A la Carte

Jaguars are naturally resourceful and adaptable hunters, which allows them to take advantage of many regional 'delicacies'. Although they're not known as man-eaters, they prey on a vast range of species. Close to villages and towns, jaguars may be opportunistic, even lazy, hunters, scavenging for food, as well as taking advantage of the relatively easy pickings to be found on farms. They're especially fond of cattle and horses, which often brings them into conflict with the human population. In the forests, jaguars prove themselves to be every inch the agile cat. They are skilled climbers and regularly hunt for monkeys and reptiles in the low branches – as well as stalking more land-based prey such as the capybara. In swamps, they think nothing of tackling a caiman in its own element. If any place can be called home for such a widely travelled and flexible predator, it's these marsh and swamp regions. Jaguars are strong swimmers and prefer a residence close to the water's edge. Here they have the best of both worlds: lots of easily available prey, such as fish and turtles, plus a wide range of larger animals, who come to the water's edge to drink.

Top Cat

In appearance, jaguars can vary significantly from region to region, which is another reason for their great success as predators. All jaguars have fur with rosettes of black spots on a lighter background. In densely forested areas, however, they're darker in colour, which gives them better camouflage in the lower light beneath the tree canopy. They also tend to be much smaller than those found in open scrub lands, which makes it easier for them to hide from both predators and prey.

Jaguars are also able to adapt their behaviour to suit their environment. In areas close to human habitation, they are mainly nocturnal, using their excellent eyesight, hearing and sense of smell to hunt under the cover of darkness. In other, more rural regions, they are diurnal or crepuscular, operating in that strange, twilight world around sunset or before dawn. Champion predators need to be flexible, and jaguars have gained their place at the top of the South and Central American food chain by being just that.

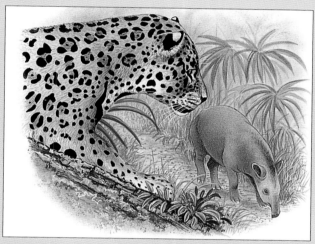

The jaguar patently waits in the darkness on a low branch for the grazing tapir to wander underneath.

The tapir is knocked to the ground as the jaguar pounces, gripping with its claws to deliver a bite to the throat.

After the tapir dies from blood loss and shock, the jaguar drags it to cover before devouring the guts.

After eating, the jaguar hides the carcass from scavengers under earth and leaves, ready for his next meal.

Maned Wolf

Maned Wolves may not be Latin America's largest or most fearsome predator, but, like all wolves, they're stealthy, fast and cunning hunters. Equipped with sharp teeth for tearing meat, and claws to dig up burrowing prey, the Maned Wolf is a skilled and highly specialized killer on the grasslands and plains.

Key Facts	ORDER *Carnivora* / FAMILY *Canidae* GENUS & SPECIES *Chrysocyon brachyurus*
Weight	20–25kg (44–55lb)
Height	74–87cm (29–34in)
Length Head & Body Tail	1.25–1.3m (4ft 1in–4ft 3in) 28–45cm (11–18in)
Sexual Maturity	1 year, but rarely breeds until second year
Mating season	April to July, peaking in May and June
Gestation period	62–66 days
Number of young	2 to 5
Breeding interval	1 year
Typical diet	Small rodents, birds, insects, fruit and other plant matter
Lifespan	13 years in captivity

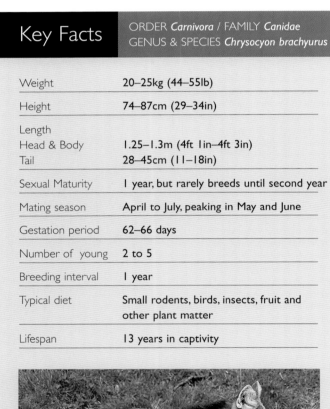

Teeth

Small, sharp carnassial teeth are used for shredding and slicing up skin and muscle, while broad, heavy molars grind and pulp vegetable matter.

Feet

Walking on just the tips of its toes, a Maned Wolf can move swiftly and silently. By spreading these toes out, it can also avoid sinking in mud.

Maned Wolf habitats

Maned Wolves were once common in Central and South America. Numbers of wild wolves are now falling rapidly, although there are captive breeding programmes in both North America and Australia.

Ancient Ancestors
Maned Wolves are perhaps among the most attractive of South America's wild inhabitants. Standing around 1m (3ft 3in) tall, this striking member of the family *Canidae* (dog family) looks at first glance like a fox on stilts. With reddish fur, a long bushy tail and large ears, this beautiful predator is made all the more noticeable by its incredibly long legs. It is believed that the Maned Wolf is descended from wild dogs and jackals which arrived on the American continent many thousands of years ago from Africa. Over time, they became adapted to a life on the South American pampas (grassy plains). For example, it has splayed feet, which are ideal in wet and marshy regions, while its long legs enable it to see easily over the tall grasses. This also gives it a pacing gait that allows it to travel long distances with ease. This is important during the dry season when prey may be scarce and the wolf has to travel further afield to feed.

Talk to Me
A Maned Wolf's range may cover up to 64km (40 miles), which it shares with its mate. Wolves are pack animals and are instinctively sociable, but while they mate for life, males and females spend very little time together. Instead, they leave each other scent-laden messages. This may provide information about recent kills, but it will also include chemicals that tell the male when the female is ready to mate again.

After mating, which occurs between August and October, the male and female go their separate ways again. The female will make a 'nest' above ground in thick brush, and between 62 and 66 days later a litter of up to 5 cubs is born. It's the female's job to prepare these youngsters for a life on the grasslands. It will take around a year for the cubs to fully mature, but once grown, they'll form small family packs with their parents until they're ready to set off on their own. Maned Wolves are highly vocal and use calls as well as scent to keep in touch with other pack members. A whine is usually a call for help from an

Using its acute senses, the Maned Wolf looks, smells and listens for potential prey in the tall grass.

The wolf hones in on any sound that might mean food, however faint. It stands still while it concentrates.

Comparisons

Although it looks more doglike, the Colpeo Fox is closely related to its long-legged relative. Colpeos live high in the Andes and so have thick coats to keep out the cold. Sadly, this lush coat has made the Colpeo Fox a favourite with hunters and the animal is now becoming endangered.

Colpeo Fox

Maned Wolf

injured wolf. A growl is territorial: it means, Stay away. A full-blooded howl is generally used to assemble other pack members together but sometimes wolves, just like the domestic dog, will howl for the sheer pleasure of it.

Meat and Fruit

During the day, the Maned Wolf prefers to take it easy, sleeping in the sub-tropical sun. It's at night that the wolf turns hunter. Anything from field mice to birds, fish and lizards is acceptable food to a hungry wolf, but guinea pigs, chicken and the occasional lamb are a favourite too. Maned Wolves are cunning hunters, using natural camouflage to hide their movements until they are ready

to pounce. When they do, the kill is fast and furious, and prey is usually devoured on the spot. Unusually for members of the dog family, though, Maned Wolves also supplement their diet with fruit, and have special grinding teeth (molars) for this purpose.

There are many Latin American superstitions about Maned Wolves. One says that an eye, freshly plucked from a live wolf, will bring good luck to gamblers. Another says that Maned Wolves have snakes living in their stomachs. Actually, they do have large 'kidney worms', a type of parasite, and it is possible that one of the fruits the wolf is most fond of, solanum lycocarpum, has medicinal or pain-killing properties.

To add the element of surprise, the Maned Wolf attacks like a fox, leaping on its prey with all feet off the ground.

Its lean forepaws pin the prey down before killing it with a bite. The wolf devours its meal before continuing on.

Mexican Red-Kneed Spider

The Mexican Red-Kneed Spider might not sound dangerous, but if we were to call it by its more familiar tag of the Red-Kneed Tarantula, many of us would feel a distinct chill at the mere mention of its name. Yet, what's in a name? When it comes to human fears, it seems that sometimes a bad reputation is all that's needed to be labelled as 'dangerous'.

Fangs
Tarantulas are unable to digest food inside their body. Instead, they inject their victim with a chemical cocktail containing digestive juices. Some will additionally inject their victim with poison.

Key Facts

ORDER *Araneae* / FAMILY *Theraphosidae* / GENUS & SPECIES *Brachypelma smithi*

Weight	57g (2oz)
Body length	7.5–10cm (3–4in)
Legspan	Up to 18cm (7in)
Sexual maturity	5–7 years
Number of eggs	Up to 400
Hatching time	2 to 3 months
Birth interval	1 year
Typical diet	Large insects; possibly small lizards and small rodents
Lifespan	Up to 30 years

Spinnerets

Specialized organs on the spider's abdomen produce silk. This is variously used to build nests or to help the spider to hold its prey.

Eyes

Most spiders have 6 or 8 eyes. The Red-Kneed Spider has 8, and these are arranged in pairs on a raised dome on top of its head.

Originally, the name 'tarantula' referred to a type of Wolf Spider found in Italy. According to legend, the poison from this spider's bite would make you run around and shout in a wild and uncontrolled way. The only cure was to dance enough to 'work' the poison out of the body. This became the basis for the popular Italian folk dance known as the tarantella. Today the name tarantula is usually given specifically to larger, hairy spiders of the type found in tropical regions, especially Mexico and South America.

Dangerous…

South America is home to both the world's smallest and largest species of spiders. Tarantulas are officially the largest, and within this order the Goliath Bird-Eating Tarantula of Guyana can grow to an incredible 28cm (11in) in length, including legs. Mexican Red-Kneed Spiders generally reach a more modest 15–18cm (6–7in).

Red-kneed spiders are hunting spiders, which means that they're fast and aggressive predators. They don't build webs but instead track, stalk and then run down their prey. When threatened, a Mexican tarantula may shed 'urticating' hairs, which can cause a burning sensation, rashes and even blindness if they get into the eye. At the front of their head, they also have two sharp fangs, which can be used to inject their victim with poison. These adaptations make them dangerous opponents. Within their own environment, they are undoubtedly formidable hunters, but much of our reaction to tarantulas is based on phobias and misinformation.

But to Whom?

It may seem strange that animals which are relatively small and harmless, compared to man, should generate such animosity and aggression. However, surveys of 'most hated

The Red-Kneed Spider lies in wait, using the hairs on its legs to feel for the vibrations of approaching prey.

A grasshopper lands nearby. Alerted to the new arrival, the spider pounces on its hapless victim.

The spider's fangs deliver the paralysing venom and soon the grasshopper stops struggling.

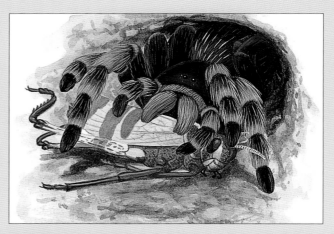

Using its jaws to open the grasshopper up, the spider then sucks out the body juices.

animals' are consistently topped by spiders, narrowly beating snakes. Fear of spiders is called arachnophobia and it's estimated that, worldwide, 50 per cent of women and 10 per cent of men have some form of spider phobia. This can be anything from a mild discomfort to physical reactions like sweating, dizziness and vomiting. No one knows why such reactions may occur, although some psychologists have suggested that many of our fears are based on ancient survival instincts. In danger situations, all animals have a fight of flight response. It may simply be that phobias are a form of early warning system.

In studies, many phobias are successfully treated by regular exposure to the object of fear. It's also known that native people who live in regions where tarantulas are common aren't afraid of them. In fact, they consider them to be a tasty delicacy! So the fear many of us experience may be based not on what we know, but on what we don't know.

A Liquid Meal
Amazingly, tarantulas are unable to digest food inside their own body. Instead all the complex chemistry involved in breaking down a food's fats, proteins, carbohydrates, vitamins and minerals takes place inside the victim's body. Tarantulas do this by injecting their victim with a cocktail of digestive juices, which are actually acids that break down the prey's flesh into a more manageable form.

Mexican Red-Kneed Spider habitats

Spiders live on a liquid diet. Once the digestive fluids have 'pre-digested' their prey, they use a short strawlike appendage to pierce its skin and suck up their victim's body, like soup. Mexican Red-Kneed Spiders live on a varied diet of insects, plus the occasional small reptile, amphibian or mammal. It's possible for a large tarantula to reduce a mouse to fur and bones in just 36 hours!

Comparisons

Tarantulas come in all shapes and sizes and some can grow to quite alarming sizes. The largest is the Goliath Bird-Eating Spider. This monster of the spider world can be found in Brazil, Trinidad and Guyana, where it preys on small lizards, snakes and frogs.

Red-Kneed Spider   Avicularia   Goliath Spider

Piranha

Few fish apart from the shark have made it into the public consciousness, but the piranha has managed to carve itself a grizzly reputation. The subject of films, books and legend, the piranha has become one of South America's most infamous river residents, but the question has to be asked: can a small fish really be dangerous?

Lake Piranha   Black Piranha   White Piranha

Size and shape

There are around 18 species of piranha, which are found throughout lakes and rivers in South America. These vary greatly in size but, as these pictures show, most have flat, broad and muscular bodies.

Key Facts	ORDER *Characiformes* / FAMILY *Characidae* GENUS & SPECIES *Various*
Weight	1–2kg (2lb 3oz–4lb 6oz)
Length	15–60cm (6–24in)
Sexual maturity	1–2 years
Breeding season	Onset of the wet season – December to May, depending on location
Number of eggs	From a few hundred up to 5000
Breeding interval	1 year
Typical diet	Mainly fish, but also mammals, birds, insects and fallen seeds, fruit and leaves; some species are vegetarian
Lifespan	Up to 5 years

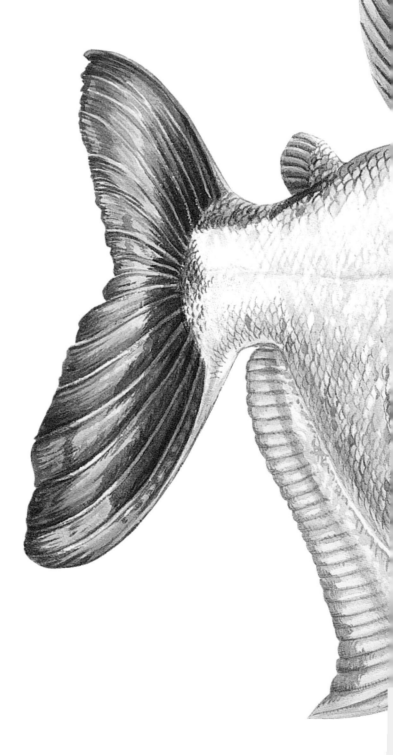

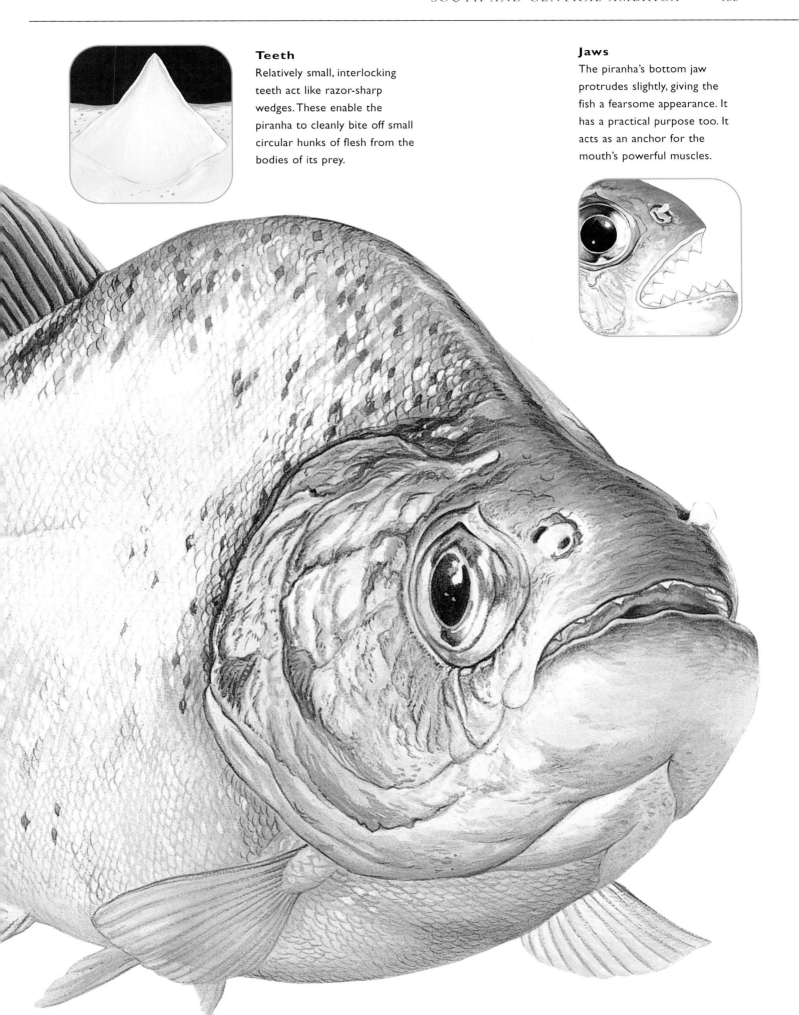

Teeth
Relatively small, interlocking teeth act like razor-sharp wedges. These enable the piranha to cleanly bite off small circular hunks of flesh from the bodies of its prey.

Jaws
The piranha's bottom jaw protrudes slightly, giving the fish a fearsome appearance. It has a practical purpose too. It acts as an anchor for the mouth's powerful muscles.

Piranha habitats

The name 'piranha' refers to around 18 species of fish found throughout South America. These carnivorous fish vary in length from the Red Piranha in eastern Brazil, which grows to 60cm (24in), to the Red Piranha, which at 30cm (almost 12in) is more like the average size.

Fearsome Fathers

In the early 2000s, the world's attention turned to the town of Santa Cruz da Conceicao in Brazil after an unprecedented series of attacks by piranhas on bathers in the local river. While stories of man-eating piranhas abound in South America, these attacks were unusual in that they were not just well documented, but were also frequent and ferocious.

A piranha's teeth are relatively small, but incredibly sharp and interlocking. When they attack, their strong jaws do most of the damage, closing these flat, razor-sharp wedges together with such force that they slice off small circular hunks of flesh, much like taking a bite out of a sandwich. Over five weekends in summer 2002, a total of 50 attacks were recorded in Santa Cruz – with many of the piranhas' victims losing fingers and toes.

Eventually the reason was discovered: the recent damming of the Rio Mogi Guacu. Piranhas lay their larvae (eggs) in waterweeds that collect in slow-moving water. So, in building the dam, the engineers had inadvertently created the perfect piranha nursery. While the problem for the bathers may have been the piranha, the problem as far as the fish were concerned was the humans. Male piranhas guard their eggs and young (fry) until they are able to fend for themselves, so what seemed like predatory behaviour to the inhabitants of Santa Cruz was is, in fact, good piranha parenting.

Pack Attack!

Usually piranhas are lone predators, but they will gather together in huge shoals (groups) when conditions are right. Although highly fanciful travellers' tales have been told about piranhas, it's no exaggeration to say that a shoal of piranhas can be as effective hunters as a pack of wolves.

Comparisons

Looking at the Hawaiian Sabre-Toothed Blenny, you'd be forgiven for thinking that this carnivorous fish is just as predatory as the piranha. While it is a meat-eater, however, those incredible fangs are defensive only. It prefers to eat worms and crustaceans.

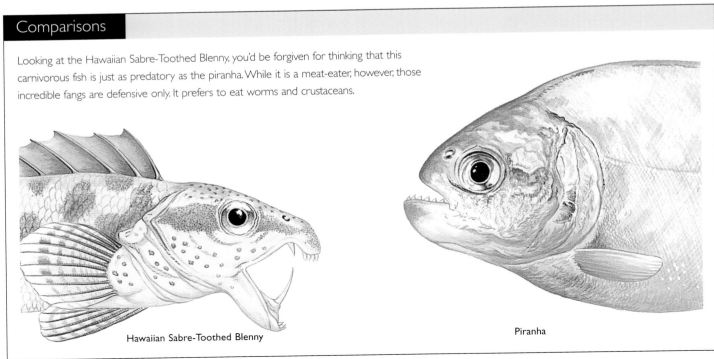

Hawaiian Sabre-Toothed Blenny

Piranha

It used to be believed that piranhas were attracted by the smell of blood, but any unusual movement in the water may be a trigger. Working together, a shoal of piranhas, which may be several thousand strong, makes an efficient disposal unit. They have a reputation as ravenous carnivores that will strip to the bone any animal entering the water in a matter of minutes. In fact, a swarm of piranha has actually been recorded eating a fully grown 45kg (100lb) pig in less than a minute.

Unbalancing the Scales

Piranhas are such voracious predators that they present a serious threat to any ecosystem other than their own. In ordinary circumstances, piranhas feed mainly on other fish, plus the occasional seed or piece of fruit that falls into the water. As an attractive fish, however, they used to be a very popular pet, and were exported and sold to enthusiasts throughout the world. Unfortunately, when collectors tired of them, the fish were often simply released into local rivers and streams. A healthy ecosystem is one in balance: there are enough prey animals for the predators to thrive and just enough predators not to exhaust the supply of prey animals. When alien species are introduced into an ecosystem, the balance is thrown. Native species have no natural defences against the predator, which quickly works its way through the resources available. This has happened in numerous areas where piranhas have been released, to the point that today even zoos have to apply for licences to move these potentially devastating carnivores into any new environment.

Sensitive to unusual vibrations carried through the water, the shoal of piranas are on the lookout for prey.

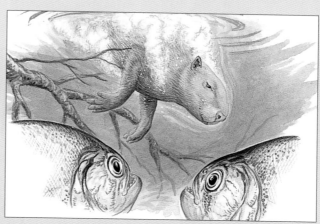

From below, the piranas see a capybara plunge into the water, probably fleeing from another predator.

Usually the piranas will not attack a large mammal, but they are drawn to the splashing as the capybara swims to the other bank.

The attack becomes increasingly frenzied as more blood enters the water. Each fish attacks, then retreats to swallow, and then attacks again.

Strawberry Poison-Arrow Frog

The skin of a poison-arrow frog is loaded with toxins, but these are not used for hunting, they're purely protective. If ingested, however, these poisons can cause anything from paralysis to death. So, if you're a rainforest inhabitant, the Strawberry Poison-Arrow Frog may be the most dangerous meal you'll ever have.

Mouth and Tongue

Using its long, sticky tongue like an additional hand enables the Strawberry Poison-Arrow Frog to catch and hold prey. Its flexible jaws then open to swallow prey whole.

Key Facts	ORDER *Anura* / FAMILY *Dendrobatidae* GENUS & SPECIES *Dendrobates pumilio*
Length	2.5cm (1in)
Mating season	All year
Number of eggs	4–6; tadpoles develop in tiny pools of water in tree-growing plants, where they are fed unfertilized eggs
Incubation period	Varies according to the temperature
Birth interval	12 months
Typical diet	Ants; other small insects and spiders
Lifespan	3–5 years

Feet and pads

At the end of each digit are round suckers, which allows the poison-arrow frog to cling to wet and glossy surfaces.

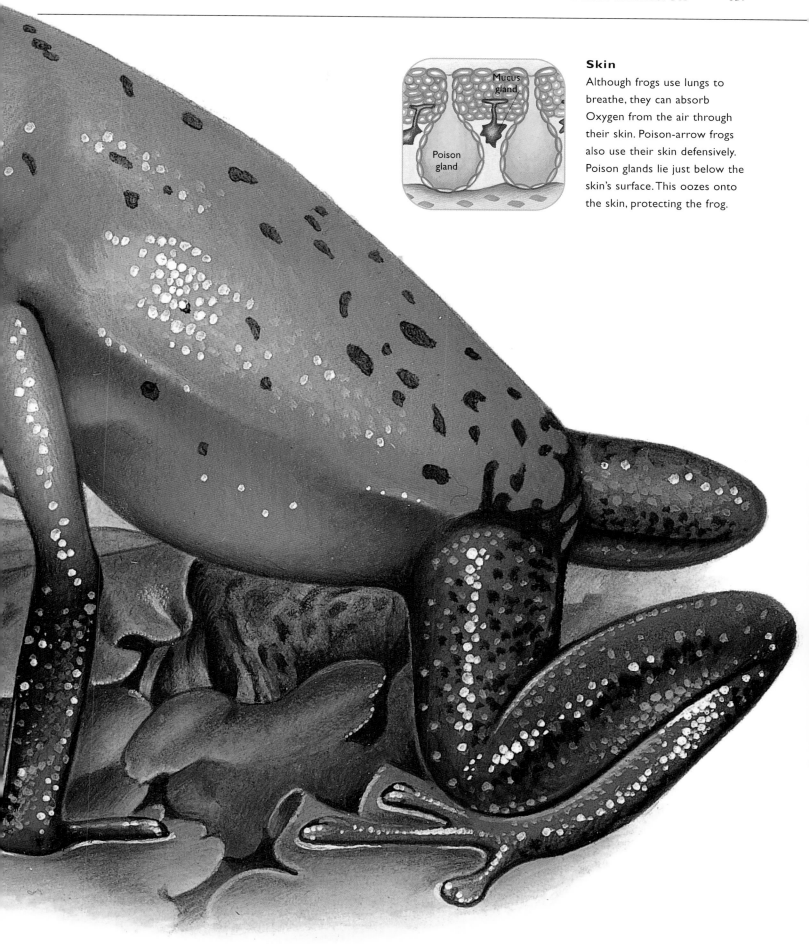

Skin

Although frogs use lungs to breathe, they can absorb Oxygen from the air through their skin. Poison-arrow frogs also use their skin defensively. Poison glands lie just below the skin's surface. This oozes onto the skin, protecting the frog.

Comparisons

Dendrobates auratus

D. lehmani

D. terribilis

D. azureus

It's estimated that there are as many as 100 species of poison-arrow or poison-dart frogs. No one knows for sure. These spectacularly coloured members of the *Dendrobates* Genus make their homes amongst the rainforests of South and Central America. Unlike frogs living in cold climates, which tend to hibernate during the winter months, tropical poison-arrow frogs tend to be active all year round. As their name implies, all varieties of poison-arrow frog – regardless of their colour – produce poison, which is secreted from glands that are located just beneath the frog's skin. They share similar mating and breeding habits too.

Poison-arrow frogs can be found in a spectrum of fantastical colours. The strawberry version, as the name suggests, is a vibrant red, with dark blue legs, which also give it the alternative name of the 'blue jean' frog. Yet in Panama, strawberry frogs may also be orange with black spots, red with black spots, or red with white spots. Other types of poison-arrow frogs come in a similarly vivid variety of solid colours, such as blue or orange, and there are also spotted alternatives.

Health Warning

Flashes of bright red beneath the rainforest canopy tell visitors that a Strawberry Poison-Arrow Frog is near. Many animals work hard to make themselves blend in with the environment, but for members of the poison-arrow (or poison-dart) family safety is dependent on being easily visible.

The reason for such an array of colours isn't camouflage. The frog is using skin pigment to tell the world to stay well away – and any animal unlucky enough to ignore this crystal-clear message is in for a nasty surprise. Many animals use chemicals defensively. Skunks spray attackers with foul smelling vapours. Cobras spit venom at their enemy's eyes. The toxins on the Strawberry Poison-Arrow Frog's skin can be a deadly mouthful. Animals quickly learn to recognize 'dangerous' animals, especially if they have such bright coloration, and learn to keep their distance.

Slow and Steady Wins the Race

As predators, frogs are unusual: they don't rely on speed or aggression. They are simply patient hunters. Arrow frogs live amongst the damp leaf litter on the rainforest floor. Here one will sit almost stock still until prey approaches, and then shoot out its long, sticky tongue to catch and pull a meal to its mouth.

Adult strawberry frogs eat a wide range of insects, and occasionally small invertebrates and reptiles. Unlike most amphibians, which often lay batches of several hundred eggs at a time, the strawberry frog lays no more than five

Strawberry Poison-Arrow Frog habitats

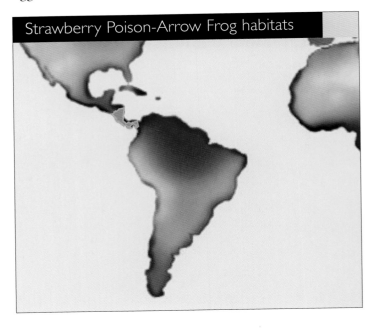

eggs. When they hatch, the young frogs (tadpoles) are carried on the parent's back to small pools of water that collect in the hollows of large bromeliad leaves. It's important for the health of the babies that they are kept, not just damp, but also separate from each other since the hungry young may become cannibalistic. Parent strawberry frogs go to great lengths to secure safe and suitable 'nests'. As the mother lays so few eggs, she feeds the tadpoles with the remaining unfertilized eggs. This rich food source is important, as it gives the tadpoles the energy they need to change (metamorphose) into frogs. Tadpoles are quite different from adult frogs. They have no legs, a tail, and breathe through gills like fish. Over several months, the tadpoles slowly grow legs, lose their tail and develop lungs. At this point, they are ready to start hunting for themselves.

Deadly!

Although poison-arrow frogs don't use their powerful toxin for hunting, other animals do – namely humans. The poison from an arrow frog varies in strength between one variety and another. One, for example, secretes a toxin that is among the most powerful known to science. The poison it produces, called batrachotoxin, is actually ten times worse than tetrodotoxin, the deadly toxin produced by the Puffer Fish.

In the jungles of South America, native tribespeople have used frog toxins for centuries, daubing them on the tips of their darts, which is how this species of frog gained its usual name. Just one dart carries enough poison to kill a small monkey, and even a tiny poison-arrow frog has enough toxin on its skin for 50 darts. This small amphibian is quite an impressive killer after all!

The female chooses a damp, shady spot to lay her eggs. After he has fertilized them, the male keeps the eggs moist with water from his skin.

Once the eggs have hatched, the tadpole climbs onto the female's back and is then held there by the wet and sticky mucus she secretes.

In the treetops, the tadpole is taken to a small nursery pool. The female lays an egg on which the tadpole can feed.

The rich diet of the egg allows the tadpole to grow quickly into a tiny frog less than 1cm (³/8in) long.

NORTH
ATLANTIC
OCEAN

Scandinavia

NORTH
SEA

EUROPE

Carpathian Mts.

Alps

Massif
Central

Apennines

BLACK
SEA

Caucasus

MEDITERRANEAN
SEA

North Africa

Europe

Europe comprises just a fifth of the huge 'Eurasian' landmass, but embraces an astonishing variety of peoples and cultures. Despite being often densly populated and industrialized, the many different regions of European still contain their fair share of dangerous animals.

~

If we can imagine viewing Europe from the air, what we'd see would seem like some huge, intricate, 3-D jigsaw. Within what is a relatively small continent – the smallest apart from Australasia, in fact – we would nevertheless find some startling differences. Travelling to the far northwest, we would find Iceland. Located on the very edge of Europe, this icy landmass, in the heart of the wind-blown, grey Atlantic Ocean, just clips the rim of the Arctic Circle. Moving directly due south, we reach the British Isles, a small, green, temperate island group containing four highly individual and historic countries (Ireland, and the United Kingdom of England, Wales, Scotland and Northern Ireland). Turning into mainland Europe, the contrasts continue. To the south, we would find the dry, parched plains of Spain and Portugal. To the east, the rolling vineyards of France. Travelling faster, now, we advance through the European heartland, crossing border after border, through flower-strewn, Alpine mountain villages, and past rich, flat farmland, until we finally reach the sweeping expanse of the Russian steppes.

Bound together by common historical and cultural links, the countries through which we have just made our virtual tour remain nevertheless separate pieces within the giant European jigsaw. Each contains its own unique geology, flora and fauna. While many of Europe's larger predators have long since been driven out by the growth of cities and industry, each nation state still offers a home to its very own 'dangerous' animals.

Badger

Europe's own mini striped badger makes a surprising addition to the roll-call of dangerous animals. Yet just one look at their canine teeth, powerful forepaws and razor-sharp claws tells us all we need to know about these striking-looking carnivores.

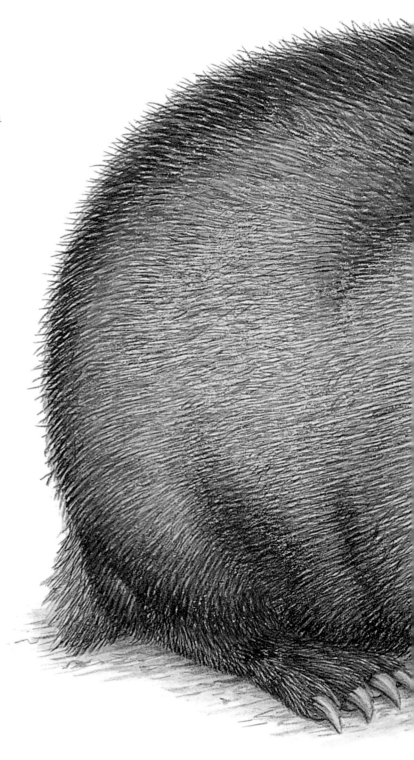

Key Facts	ORDER *Carnivora* / FAMILY *Mustelidae* GENUS & SPECIES *Meles meles*
Weight	10–16kg (22–35lb)
Length Head & body Tail	56–90cm (22–35in) 12–20cm (4³/4–7³/4in)
Sexual maturity	1 year
Mating season	All year
Gestation period	6–8 weeks, not including variable delayed implantation
Number of young	2 to 6, but usually 3 or 4
Birth interval	1 year
Typical diet	Earthworms, small mammals, insects, fruit, plants, cereals and roots
Lifespan	7–10 years

Mammae

To feed her young, the female Eurasian Badger is equipped with 2 pairs of mammae. These are the organs which produce the milk.

Jaws
Badgers have large jaws, powerful jaw muscles and sharp, deeply rooted teeth.

Claws
The claws on the Eurasian Badger's fore-feet are around twice the length as those on their hind-feet. They are used defensively as well as to dig.

Badger habitats

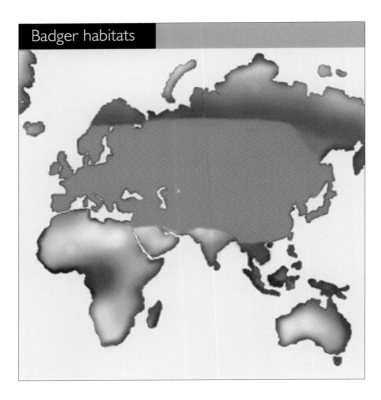

Badgers can be found all over the world. From the tiny Burmese Ferret Badger to the stubby legged American Badger, these beautiful mammals have suffered much at the hands of humans, but they are now protected in many parts of Europe.

Stripy and Stinky

Badgers are short and stockily built but, although they look very like small bears, they're actually members of the family *Mustelidae*, which includes stoats and weasels. All

members of this family produce a strongly scented 'musk' – hence the name. This is used as a form of communication, giving messages about potential threats, marking territory and attracting mates. The potency of these smells varies, but it's no surprise to learn that Stink Badgers well deserve the name!

Eurasian Badgers are the largest of the badger species and can weigh as much as 20kg (44lb), although 9–11kg (20–24 lb) is average. It is the badgers' coloration that most people will be familiar with. European Badgers have a very distinctive and prominent black and white pattern on their faces, which makes them look striped. Their body fur is generally grizzled (greyish) with long, tough hairs that have been popular for use in shaving brushes for centuries. This bold insignia seems to be a purely European feature: American Badgers are a much paler brown, with very light, barely visible face stripes, while Ferret Badgers have no stripes at all.

Underground Comfort

Badgers make their homes in wooded areas. They're generally nocturnal and spend much of the daylight hours underground in a burrow called a set. Occasionally, though, they can be seen during the day, sometimes resting in open-air nests. A badger's forelegs are particularly well developed and tipped with long, sharp claws. This makes them excellent excavators. The sets of Eurasian Badgers can be extremely elaborate constructions that have been used, and 'improved', by generations. These underground penthouses may have up to 20 separate entrances, with straw-lined chambers for sleeping or nursing, all connected by metres of winding passages. Badgers are fastidious

Comparisons

Members of the Family *Mustelidae*, which includes weasels, otters skunks and wolverines, tend to be fairly prevalent throughout the world's Temperate regions (areas with a mild climate). Eurasian Badgers make their homes across much of Europe and Central Asia. The American Badger is similarly widespread and can be found as far afield as Canada, Northern and Central America and Southern Mexico. As well as covering a huge range, the American Badger shares many other 'Mustelidae' features. They're fast and active predators, with slender bodies, short legs and small heads.

American Badger

Eurasian Badger

housekeepers and allow no refuse within the set, building separate toilet areas above ground.

Where food is plentiful, badgers are social animals who live in groups, called 'clans', of up to 12. These groups are headed by a dominant male (boar) and female (sow). In less prosperous regions, males are solitary and will fight fiercely to keep other males from their territory.

Full of Surprises

A European Badger's favourite natural food is worms, which form 60 per cent of its diet. They regularly add seeds, nuts, berries and honey to the menu too. Just because badgers have a taste for invertebrates, however, doesn't mean that they lack hunting skills. They're powerful predators and regularly kill small animals such as rabbits, frogs or ground-nesting birds for food. Honey

Badgers have even been known to tackle young antelope. Yet badgers seem to prefer the easy life. They're often found around areas of human habitation, where scavenging can result in a free meal. With a well-developed sense of smell, carrion is a favourite. In fact, badgers in Asia occasionally dig up graves and eat human corpses! Luckily, they tend to avoid live humans, although in England, in 2003, a rogue badger attacked and seriously injured five people before it was caught. This story is a good demonstration of just how dangerous badgers can be, a fact that our ancestors knew well. In the Middle Ages, badger-baiting was a popular, if exceptionally cruel, sport in Europe. Badgers would be dug up from their sets and forced to fight dogs. To 'even up' the odds, the badgers would often have their jaws or paws broken; even so, they would usually maul the dog severely.

The badger will regularly refresh its predominantly grass bedding by bringing it out into the open.

With precision and care, the badger spreads the bedding out to allow it to air and dry.

Once the badger is satisfied the bedding is dry, it collects the grass under its chin and, bit by bit, brings it all back inside.

The badger lies down to sleep on the refreshed bedding once it has all been brought back into the sett.

Black Rat

Originating in Asia, the Black Rat has hitch-hiked its way across the world, bringing death and disease in its wake. These small, agile mammals may look inoffensive but they're host to a wealth of deadly diseases, including bubonic plague, typhus and rabies. This makes them, albeit indirectly, one of the world's biggest killers.

Key Facts	ORDER *Rodentiaa* / FAMILY *Muridae* GENUS & SPECIES *Rattus rattus*	
Weight	140–200g (5–7oz)	
Length Head & body Tail	Up to 23cm (9in) Up to 25cm (9³⁄₄in)	
Sexual maturity	3–5 months	
Mating season	Year round, with peaks in spring and autumn in temperate regions	
Gestation period	21 days	
Number of young	5 to 10	
Birth interval	Up to 12 litters per year	
Typical diet	Almost anything, although fruits and cereals are favoured	
Lifespan	Up to 4 years	

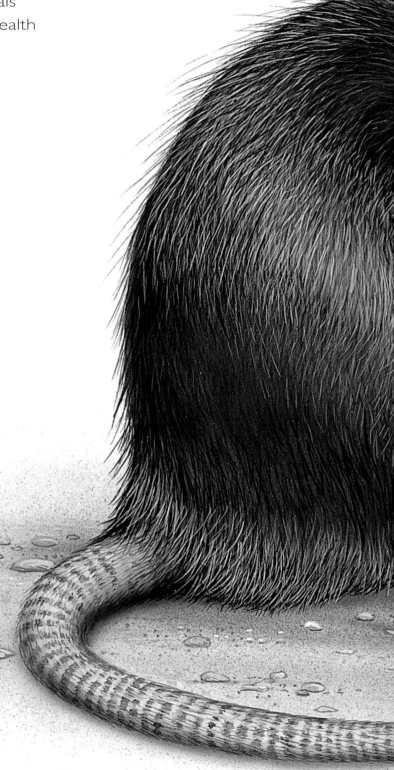

Nose and whiskers
With an incredible sense of smell, and touch-sensitive whiskers, the Black Rat has no problem finding sources of food in the dark.

Teeth
Rats' long incisor teeth grow constantly. So rats need to continually gnaw objects to keep their teeth down to a manageable length.

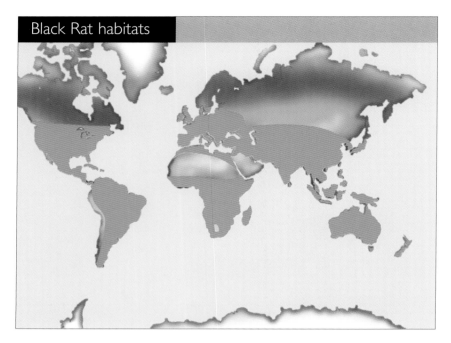

Black Rat habitats

marks on the skin, which is why this disease is also nicknamed the Black Death. Today bubonic plague can be treated by antibiotics, but left untreated it causes death in 50–90 per cent of cases in less than a week.

The results of an outbreak of plague can be devastating. One of the earliest recorded incidents killed 5000 people a day in Ancient Rome. During the fourteenth century, the Black Death killed a quarter of the entire population of Europe. As one witness commented, medieval medicine could do nothing: 'neither physicians nor medicines were effective… no one seemed to know what to do… When it took hold in a house …no one remained…even animals died.' Returning to England three centuries later, it added a further 150,000 bodies to the toll. By the nineteenth century, it was back in Asia, where it killed an additional ten million. Today there are still at least 3000 cases a year – all caused by the humble Black Rat.

Rats are one of nature's natural survivors. Found thoughout the world, these small omnivores are a highly adaptable and successful species.

The Black Death

Black Rats carry numerous deadly diseases, including bubonic plague. This virulent infection is transmitted to humans by fleas that live on the plague-carrying rats. If a flea that has bitten a diseased rat should then bite a human, the disease enters the bloodstream, causing chills and fever. As it progresses, glands under the armpits, neck and groin swell, forming buboes, which can turn into open wounds. In addition, internal bleeding causes black

Up, Up and Away

The Black Rat, or *Ratus ratus* to give it its scientific name, is now one of the rarest mammals in the UK, but still prospers in other parts of Europe. Sometimes called the 'roof rat', Black Rats tend to nest in high places. In their natural environment, in India, they dwell in trees, but since arriving in Europe they've have made the best of the man-made environment. While the Brown Rat took to the sewers, its black relative made its home amongst the

The Black Rats wait until nightfall when there is no one about before entering the grain store to feed.

The rat opens one of the grain sacks using its long, sharp teeth, allowing the grain to spill onto the floor.

Comparisons

Black Rats can come in many shades, from dark brown to jet black. Norwegian Brown Rats also come in darker tones. So the best way to tell a Black Rat from the equally prevalent brown, is that Black

Rats are usually smaller. They also look better groomed, having much sleeker coats and longer, trimmer tails.

Black Rat

Brown Rat

rafters. In contact with the human world, Black Rats present a constant danger. They have a taste for cereals and regularly damage and destroy crops and food stores. In homes, they bite through electrical cables, and leave faeces that can be picked up by children and animals. While a bite from a rat can cause rat fever or even rabies, it should be remembered that the rat is not a malicious monster: just a rather dangerous, unwanted guest.

All Bad?

Despite all the problems created by the Black Rat, there are many reasons to admire this small mammal. In the wild, they're cunning and agile hunters and may work in packs to bring down prey as large as a chicken. They're fast runners and expert climbers, using their wiry tails to help them keep their balance. In the city, they're incredibly streetwise and ever ready to exploit opportunities, especially edible ones. Like their smaller cousin, the mouse, Black Rats have become the modern world's waste disposal units. They'll eat almost anything from plastic packaging to leftover fast food. Female Black rats are dedicated mothers too. Giving birth three to six times a year, with up to ten babies at a time, a female rat spends much of her life caring for and raising young. It takes just a few months for a rat to become sexually mature, so within a year one super mother could have up to 1000 descendants. Another amazing rat achievement – but not one that's likely to endear her to the human population!

Constantly on the lookout for any possible dangers, the rat holds the grain delicately in its forepaws

Once the rat is sated, it scatters some of the grain before marking it as its own with urine.

Lammergeier

To a Spanish farmer staring up at the threatening silhouette of a Lammergeier, Europe's largest bird of prey must seem like one of the giant raptors of ancient mythology. With a reputation as large as its size, the truth about just how dangerous this incredible animal is, can be hard to determine.

Juvenile

In common with most birds, juvenile (young) Lammergeiers have a different colouring to the adults. Before they acquire their full adult plumage, their body feathers tend to be considerably darker.

Bill

The Lammergeier's bill is almost as long as its entire head, measuring 8cm (3in) from base to tip. Covering much of this bill are fine, dark feathers.

Claws

Lammergeiers have a reputation for carrying off live prey. Such long, sharp claws are certainly ideal for the job.

Key Facts	ORDER *Falconiformes* / FAMILY *Accipitridae* GENUS & SPECIES *Gypaetus barbatus*
Weight	4.5–7kg (10lb–15lb 6 oz)
Length	1–1.15m (3ft 3in–3ft 8in)
Wingspan	2.65–2.82m (8ft 7in–9ft 3in)
Sexual maturity	5 years
Breeding season	Varies according to region; January to July in southern Europe
Number of eggs	Usually 1 or 2, occasionally 3
Incubation period	55–60 days
Fledging period	100–110 days
Breeding interval	1 year
Typical diet	Hunts small mammals and birds; carrion
Lifespan	Unknown

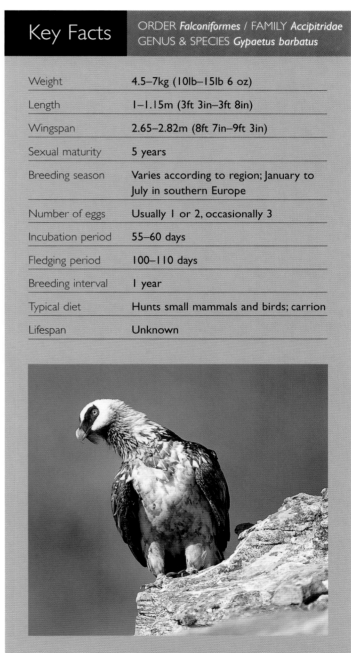

Lammergeier habitats

The Lammergeier is found mainly in mountainous regions of mainland Europe, particularly Spain, although sub-species range as far afield as Africa and Asia.

A Case of Mistaken Identity …

With long, featherless necks, pinched, fleshy faces and a crouched stance, vultures are perhaps the most unattractive and unappealing of birds. Look at a Lammergeier, and the contrast is striking. At around 1m (3ft 3in) in length, the Lammergeier is both huge and impressive. With a noble, white feathered head, tawny flank and eagle-like bearing, this elegant bird of prey is as striking as it is attractive. The surprise is that the Lammergeier and the ugly African vulture belong to the same family, as its alternative name of Bearded Vulture tells us.

Like its relatives, the Lammergeier is a scavenger, living off carrion and offal. For centuries, however, it's had a reputation for taking live prey, which it apparently kills by dropping it from a height: indeed, the Greek playwright Aeschylus (525–456 BC) is supposed to have died when a Lammergeier dropped a tortoise on his head! It has even been suggested that Lammergeier drive animals off cliffs to their deaths. This is one of the reasons given by sheep farmers for shooting this spectacular bird, since they believe that it presents a danger to their flocks. Lammergeier are certainly well equipped to deal with live prey, but there is currently little evidence to prove or disprove such tales.

Meet the Ugly Sisters

Vultures are designed to feed almost exclusively on carrion. So, when a vulture takes to the air, it's looking for corpses rather than trying to hunt. As a result, it has developed unusually long, broad wings, which are energy-saving and which allow the vulture to glide over vast distances in the search for dead or dying animals. In most birds, tail feathers are used, like the rudder on a plane, to turn. However, as vultures don't need to be particularly

Comparisons

Although they're unlikely ever to be confused with one another, Egyptian Vultures have a similar reputation to the Lammergeier. While both birds primarily feed on carrion, Egyptian Vultures supposedly hurl foodstuffs from height to feed on their contents!

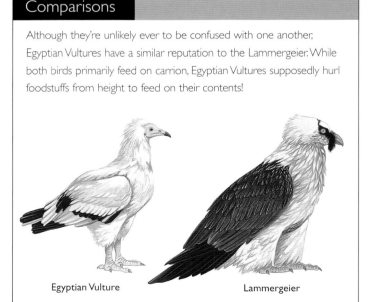

Egyptian Vulture   Lammergeier

fast or acrobatic in the air, they tend to have much shorter tails too. These adaptations can make most vultures look ungainly both in the air on the ground.

Fortunately, the vulture isn't concerned about appearances. Its most important trait is the ability to quickly and easily tear apart and digest flesh, which it can do with ease. Vultures have sharp claws, a strong beak and a huge protruding crop in their neck. This is a pouchlike part of the oesophagus (the channel that links the mouth and stomach), where food is partially digested before being swallowed. A vulture's digestive juices are powerful enough to break down bone, so little of a meal is ever wasted. The vulture's neck is especially long, allowing it to stick its whole head inside the corpse's body. As blood-matted head feathers would make it difficult for the vulture to fly, this efficient bird has also dispensed with them.

A Family Resemblance?

Lammergeiers do share some characteristics with vultures, such as their long hooked beak and sharp talons (although these are common to all birds of prey). They are, however, actually skilled aeronauts, having a long, wedge-shaped tail that enables them to glide gracefully and also turn with ease. Nor do they have the vulture's crop, but instead they drop bones from a height and then use their tongue to extract the marrow.

It is estimated that there are hundreds of species of birds of prey, including those in the family Accipitridae. This large group comprises some of the great aviators and hunters of the bird world, such as kites, hawks and eagles, which the Lammergeier so resembles. In fact when you look at the habits and characteristics of true vultures, the Lammergeier seems very much the sophisticated cousin.

The Lammergeier spots a carcass that's been abandoned by other scavangers because only the skin and the bones remain.

Taking one of the bones in its claws the Lammergeier flies off to find a suitable place to drop and break the bones.

The bird climbs to a height of about 70m (230ft) above the ground before dropping slightly to increase momentum and letting go of the bone above the rocky ground.

After picking up and dropping the bone several more times it eventually splinters on the rocks giving the Lammergeier access to the tasty marrow found inside.

Grey Wolf

Dubbed 'the children of the night' by the horror writer Bram Stoker (1847–1912), the wolf pack enjoys a dark reputation. By utilizing its strengths, as well as learning to exploit its prey's weaknesses, a Grey Wolf pack is both observant and ingenious – the ingredients for a formidable hunting machine.

Feet
Wolves, unlike wild cats, are unable to retract their claws. Instead, these are used to give additional grip when running.

Jaws
Once they have hold of their prey, Grey Wolves can exert twice the force of the jaws of a similar sized German Shepherd dog.

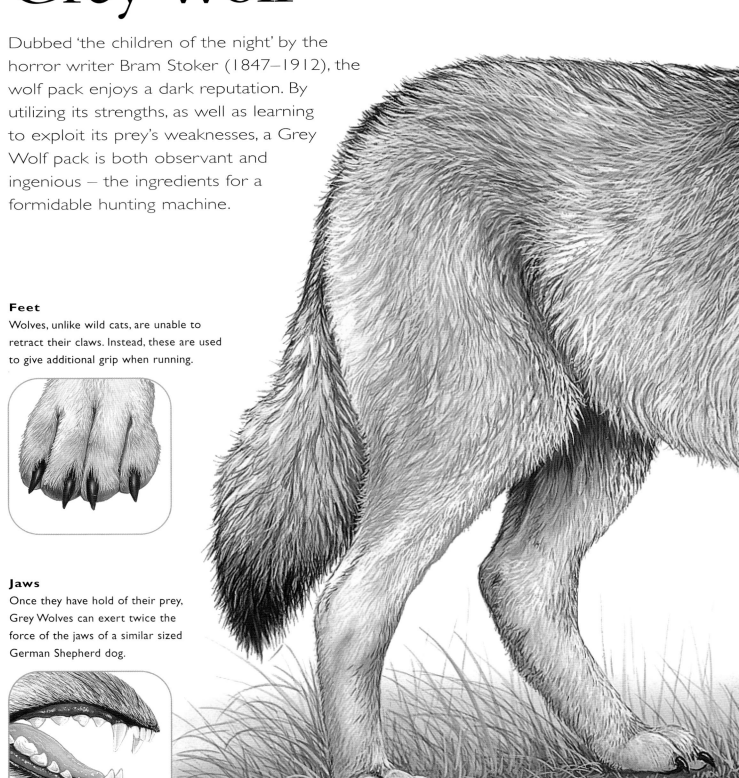

Sense of smell

Wolves' nasal passages are 4 times larger than a human's. This gives wolves an acute sense of smell. According to some estimates, this is 100 times more sensitive than a human's.

Key Facts	ORDER *Carnivora* / FAMILY *Canidae* GENUS & SPECIES *Canis lupus*
Weight	18–80kg (39–176lb)). Male heavier than female
Length Head & Body Tail	1–1.47m (3ft 3in–4ft 10in) 33–50cm (13–20in)
Shoulder height	66–96cm (26–38in)
Sexual maturity	Female 1 year; male 2 years
Mating season	Typically, January to March
Gestation period	60–63 days
Number of young	Usually 5 or 6
Birth interval	1 year
Typical diet	Mainly large grazing mammals, but some smaller mammals
Lifespan	Up to 17 years

Comparisons

Wolves come in a surprising variety of colours and sizes. While the Common European Wolf resembles a shaggy German Shepherd Dog, the Maned Wolf of Latin America has red fur and long legs – more like a fox on stilts than a wolf! Black Wolves are also fairly common, while white furred varieties can be found in the extremes of the Arctic.

| Grey Wolf | Arctic Wolf | Black Wolf | Iberian Wolf |

The scientific name for the Grey Wolf is *Canis lupus* – *canis* meaning 'dog', and *lupus* meaning 'wolf'. In fact, the Grey Wolf is the largest member of the dog family, which also includes coyotes, jackals and foxes. Although its size varies from region to region, Grey Wolves tend to be smaller in hotter climates. In Europe and North America, a male may weigh around 40kg (88lb), which makes this sub-species look strikingly similar to a German Shepherd Dog.

The Big Bad Wolf

Worldwide, Grey Wolves used to be as widespread as humans. Although they can still be found as far afield as the Arctic, North America and Russia, in Central, Northern and Eastern Europe, and from Italy to Finland, many sub-species are now endangered.

Whereas humans tamed, trained and learned to value dogs, our relationship with the wolf has always been difficult. While we may have respected them as great hunters, early humans had to compete with wolves for food – and sometimes, as fairy stories such as 'Little Red Riding Hood' remind us, we even ended up on the menu. Later, when humans started to settle in static communities and farm, wolves became a threat to our livelihood, attacking and eating farm animals. Consequently, as human settlements have spread throughout the world, the wolf population has suffered. In England, wild wolves were wiped out in the sixteenth century; in Ireland, they lasted until the nineteenth century. Today, work is just beginning to ensure that one of Europe's last great hunters is given the opportunity to survive and thrive.

Team Tactics

Wolves are pack animals. They live and hunt in family groups of usually no more than ten individuals. This cooperative existence allows them to kill much larger, faster animals than is usual for a predator of this size.

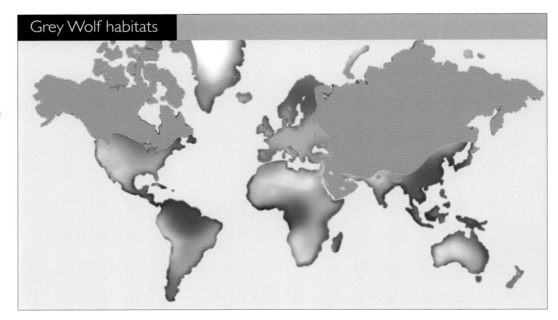

Grey Wolf habitats

Indeed, a wolf hunt is a classic exercise in teamwork. The hunt begins with the dominant male howling, to 'call' the other members of the pack together. Once gathered, they'll then set off in search of prey. Keeping downwind, so that their victim doesn't catch their scent, the wolves will inch forwards until they're close enough to try and run down their prey. Wolves have great stamina, so a chase may last for many hours until their victim tires.

However, even a kick from a tired, but enraged, reindeer or boar can kill a wolf, so the pack doesn't always take the direct approach. Instead, for example, a few members of the group may distract the prey, while others circle behind for an attack. On other occasions, the pack may divide, one half driving their prey towards the other. It's this type of incredible collaboration that has made the wolf's name a byword for skill and cunning.

Communal Living

Group living has other advantages for wolves. Within each wolf pack, there is a dominant 'alpha' male and female. The rest of the pack is usually related through several generations. Only the dominant couple will mate, but all of the pack take on the responsibility of caring for and feeding the cubs, so that more cubs are likely to survive into adulthood.

Wolf territories can be as big as 2000 square kilometres (772 square miles) in Arctic regions, so the whole pack also has responsibility for defending the range from intruders. Young wolves mature at one year old, at which age they may leave the territory to form packs of their own. Yet, if food is scarce, they may stay on for several more years, strengthening and helping the group survive during the lean times.

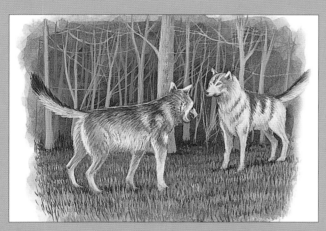

Ears pricked and tails cocked, these two male wolves attempt to establish the pecking order in the pack.

The stronger, elder male deals the younger male an aggressive bodyblow, a gesture intended to dominate.

The elder male bears his teeth. Suitably cowed by this show of dominance, the younger male lowers his ears and averts his gaze from the more senior male.

To placate the older wolf, the younger male rolls over and exposes his belly. This gesture of submissiveness shows that he accepts his junior rank.

Pike

Almost every fisherman has a tale about the 'one that got away' — that huge fish he wrestled with for a day and a half, before the line broke. Many such tales are told about the pike, which has developed an almost legendary status in the angling community for its great size, ferocity and hunting prowess.

Teeth

The pike's huge mouth has rows of razor-sharp teeth. These face backwards, giving the pike a better grip on prey.

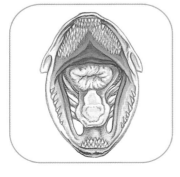

Size

Although size varies depending where in the world the pike lives, these aquatic predators can grow to immense lengths. In general, though, the female of the species tends to grow to much larger sizes.

Key Facts

ORDER *Salmoniformes* / FAMILY *Esocidae* / GENUS & SPECIES *Esox lucius*

Weight	Male 4-5kg (8lb 13oz–11lb); female up to 34kg (75lb)
Length	Male 60-90cm (24–35in); female up to 1.5m (5ft)
Sexual maturity	3–4 years
Breeding season	February–May
Breeding interval	1 year
Number of eggs	16,000 to 70,000
Typical diet	Mainly fish; also water voles, rats and waterbirds
Lifespan	Male 7-10 years; female up to 25 years

Pike distribution and habitats

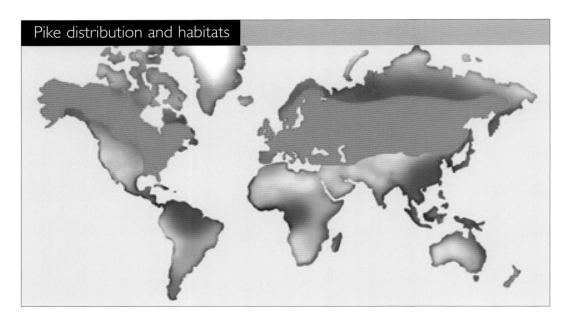

The most noticeable part of the pike is its head, which is unusually big, with an elongated, jutting beak. This is the 'business end' of this great underwater predator. Pike have only a few, but large and sharp, teeth in their bottom jaw. Their top jaw (the maxillae) is toothless, but an arsenal of backward-pointing teeth lines the palatines and vomer (roof of the mouth). Pike can open their mouths extremely wide, allowing them to indulge in a varied diet. When young, pike eat insects and small fish, but as they grow they tackle increasingly larger prey, and have a gained a reputation as voracious feeders. Certainly, they enjoy 'eating the neighbours', trout and perch in particular, but also, occasionally, other pike. They've also been known to eat aquatic birds and mammals, including moorhens, ducks, water voles and muskrats.

Pikes and Mud Minnows belong to the suborder *Esocoidei*, a group of fish widely distributed in cool, freshwater lakes and ponds throughout the northern hemisphere. The European Pike, which is known as the Northern Pike in the USA, is perhaps the most well known of this menacing-looking species, since it's a popular trophy fish and graces the walls of many fishermen's pubs and taverns.

Big Mouth, Big Appetite
The European Pike's body is long and slender. If angler's tales are to be believed, these elegant fish may reach lengths of 1.5m (almost 5ft) and weigh as much as 45kg (99lb), although they tend to be smaller in colder regions.

Armed and Deadly
The pike is well respected as one of the greatest freshwater hunters and just one look at this famed fish shows that it's well equipped for such a role. With green-grey bodies, they have excellent camouflage – one of the first and most

Comparisons

Throughout the world, nature seems to have found the same solutions to similar problems. The pike is a freshwater dweller, whereas the voracious barracuda makes its home at sea. However, both species share many features – a long, streamlined body and dorsal and anal fins close to the tail, which give the fish speed and manoeuvrability, combined with sharp, backwards facing teeth and a huge, powerful jaw. These features immediately identify both species as hunter-killers.

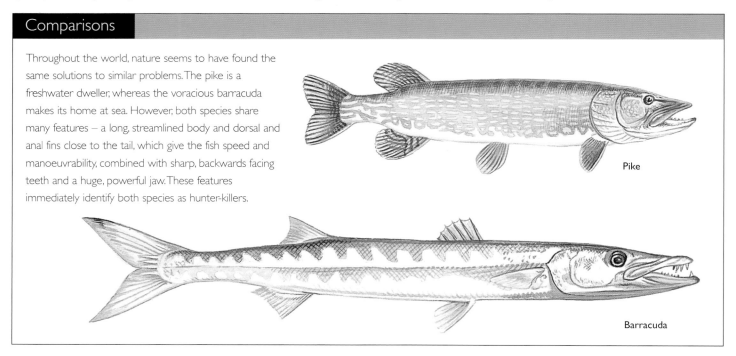

Pike

Barracuda

important requirements for a hunter. This natural coloration enables them to lie still and unseen among the reeds until a meal passes their way. On land, balance and speed are also crucial attributes for any wannabe predator. The same rules apply for fish. Fins are a fish's stabilizers, helping it to swim and keep its balance, while tails add thrust. Pikes have tightly grouped dorsal, anal and tail fins, which enable them to fine-tune their movements, as well as a long tail to propel them forwards.

When prey is within sight, the pike will use all these adaptations to make the kill. Slowly turning, the pike lunges forwards at incredible speed, using its fins and tail to great effect. Grasping its victim from below in its powerful jaws, it holds onto its prey's body, often swimming with it until it finds a safe spot to adjust its hold. Pike always swallow their victims head first, which lessens the chance of them choking. The backward-pointing palatine and vomer teeth help to guide the food down the pike's throat in a way similar to that of a snake's back-facing teeth.

Extra Credit

A final addition to the pike's hunting kit is a system of sensitive nerves called 'lateral line'. Pike have good eyesight, but the lateral line system helps them to home in on their prey. This consists of a series of tiny nerves under the skin which run along the pike's body to the head. When prey is near, these channels pick up the movement and transmit it to the brain. A change in the pattern of the lateral line indicates that the prey has changed direction. This system enables pike to hunt effectively even in the murky depths of a lake or stream.

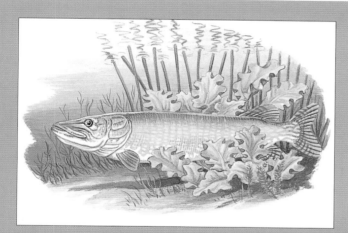

Lying motionless and camouflaged, this large pike is well hidden in the weeds as it waits for prey.

A coot, oblivious to the danger below, lands on the water. It is a tempting meal for the hidden pike.

Stealthily closing in on its target, the pike suddenly surges forward and grabs hold of the coot with its razor-sharp teeth. The pike drags its prey down into the water to drown.

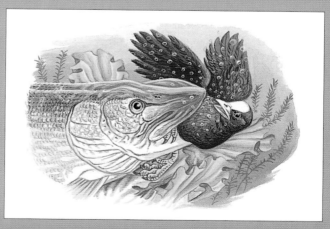

There is no escape for the coot, caught by the backward-facing teeth of the pike. When it has drowned, it is crushed and then swallowed head first.

Siberian Tiger

With its distinctive coat, great strength and agile grace, the Siberian Tiger is a truly stunning example of natural adaptation. This handsome animal was built for the kill. Unfortunately, despite being the world's largest cat, the Siberian Tiger is now more endangered than dangerous.

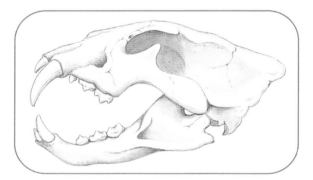

Skull and Teeth

An especially large skull anchors the Siberian Tiger's powerful jaw muscles in place, and long, canine teeth are used to stab and tear their flesh from the bone.

Foot

In common with many cats, the Siberian Tiger has soft, velvety pads on the soles of each paw. These allow it to move silently – useful when hunting.

Tongue
Covered with
hundreds of tiny
barbs, the tiger's
tongue makes a great
tool for removing
meat from the bone.

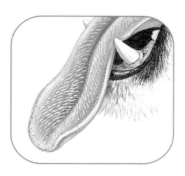

Key Facts	ORDER *Carnivora* / FAMILY *Felidae* GENUS & SPECIES *Panthera tigris altaica*
Weight	Male 180–300kg (397–662lb); female 100–165kg (221–364lb)
Length Head & Body Tail	1.6–2.8m (5ft 3in–9ft) 60–95cm (24–37in)
Shoulder Height	1–1.1m (3ft 3in–3ft 6in)
Sexual Maturity	Male 4–5 years; female 3–4 years
Mating Season	November to April
Gestation Period	104–106 days
Number of Young	1 to 6 (usually 2 or 3)
Birth Interval	2–2.5 years
Typical Diet	Pigs, deer, bears, small birds and fish
Lifespan	15 years in the wild; up to 26 in captivity

Comparisons

At just two-thirds of the average size of a Siberian Tiger, the Puma would seem to be very much the 'poor relation' of the cat family. Although small in comparison, this attractive animal shares much of its relative's natural skill and agility. Able to jump 4.5m (14ft 8in) from a standing start, and over 13.7m (45m) when running, Pumas are also powerful enough to bring down a fully grown horse.

Puma

Siberian Tiger

It is believed that all the world's tiger sub-species are descended from the Siberian Tiger. Spreading out across Europe and Asia during the Ice Age, this ancient ancestor quickly established itself as top cat in a range of environments. Today, all five of the remaining tiger sub-species – the Bengal, Indo-Chinese, Siberian, South China and Sumatran – are close to extinction. In fact, there may be only 200 Siberian Tigers left in the wild.

Making a Living…the Hard Way

When we think of tigers, we generally think of India's Bengal Tiger. This animal seems so well attuned to a life on the edges of Asia's forests and swamp lands that it's hard to see how such a huge, powerful predator could make a living in the barren, snow-bound wastes of Siberia. Yet, if we look at the behaviour of Bengal Tigers, we can see that they are natural opportunists. They prefer large prey, but will take anything that comes their way. This is a handy trait to have when food might be scarce, and Siberian Tigers have the same flexible approach to a meal. They're powerful predators, but are as likely to take an Arctic hare as a moose.

Siberia covers a huge area, around three-quarters of the whole of Russia. Much of this is ice-bound. Snow covers the ground for six months of the year, and in the far north temperatures may drop to -60°C (-7°F). Siberian Tigers prefer the warmer, forested regions of the south, but even here they need around 9kg (20lb) of meat a day to provide their bodies with enough fuel to keep warm. Siberian Tigers solve this problem by covering vast territories. Life in the north is still hard, however, since only one hunting trip in ten may be successful.

Cool for Cats

From Siberia, tigers began their epic migration across the continent. As they moved from the icy wastes into new regions, the 'original' Siberian model had to be adapted to suit the new environments and warmer climates. The Bengal Tiger, for example, has short hair and pronounced orange and black stripes on its fur. The Siberian is paler, with less obvious strips, which allows it to blend in more easily with the snow. It's the Siberian Tiger's thick shaggy coat that tells you it's a cold-climate dweller. Like all

Siberian Tiger habitats

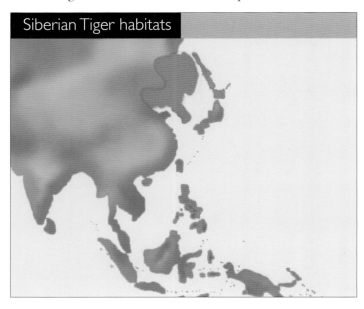

animals who live in the tundra, keeping warm is a major concern, and a long coat offers the tiger some protection against the elements.

Strangely, though, even after millions of years, tigers who make their homes in the tropics don't seem to be able to tolerate extremes of heat. They can often be seen dozing in the shade or taking a cooling dip in a nearby lake. Tigers may have adapted well to their new homes, but it seems that, at heart, they are cool customers.

Buy, Bye

Some ancient cultures regarded the tiger as a demon and believed that they could keep it from their village by making sacred offerings. Others considered it to be the 'grandfather' of the forest, who had to be treated with respect to ensure good favour. The tragedy is that, today, all a wild tiger means to some people is easy money, as tiger bones and skin are valuable commodities on the black market.

In Russia, the Siberian Tiger has received legal protection since 1992, and there are active programmes to stop such trade. Three protected areas have also been established to try to give tigers a territory where they're safe from poachers and human encroachment. About 500 Siberian Tigers now live in zoos around the world, a population that is stable and secure, but only time will tell if this wild Siberian cat ultimately joins the ranks of the now extinct Bali, Caspian and Javan Tiger.

The female finds a safe spot before giving birth, usually to two or three cubs.

The mother must carry her cubs by the scruff between resting sites, as they are helpless when newborn.

Cubs can take meat at two months and are weaned at six months, although they may starve if food is scarce.

The mother teaches the cubs how to hunt until they are two years old, by which time they strike out on their own.

Wild Boar

Mad, bad and dangerous to know, the Wild Boar is one of Europe's most aggressive wild animals. Among its own, a boar will give no quarter in fights for mating rights and dominance. To other animals, it's a neighbour who's best avoided, as this large tusked hog has a notoriously short temper.

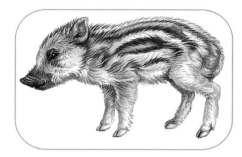

Piglet colouration

When they are first born, piglets have a distinctive striped coat. This provides ideal camouflage amongst the woods and undergrowth. These stripes slowly fade over about 6 months.

Key Facts	ORDER *Artiodactyla* / FAMILY *Suidae* GENUS & SPECIES *Sus scrofa*
Weight	50–350kg (110–772lb)
Length Head & body Tail	90–180cm (3ft–6ft) 30–40cm (12–16in))
Shoulder height	55–110cm (1ft 9in–3ft 7in)
Sexual maturity	8–10 months, but females usually breed at 18 months and males at about 5 years
Mating season	All year round in the tropics; Oct to Dec in temperate areas
Gestation period	100–140 days
Number of young	1 to 12, but usually 4 to 8
Birth interval	1 year
Typical diet	A wide range of vegetable and animal material
Lifespan	15–20 years

Split hoof

Boars are one of many animals which are described as 'cloven hoofed' (cloven means split).

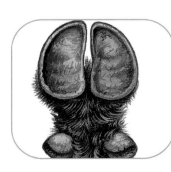

Wild Boar can be found across Europe, Africa and Asia. They've been extinct in Britain since the thirteenth century, but have recently been seen in some rural areas after escaping from farms.

Food, Glorious Food

Living in lairs in quiet wooded areas, boars are natural foragers who know how to make the best use of the resources around them. A Wild Boar's best asset is its snout – that long, square nose, which it uses as a multi-purpose tool to sniff out and dig up roots, tubers, acorns, chestnuts and earthworms. They eat carrion too, especially in the winter, and occasionally kill small birds, mammals and reptiles. Recently, boars have also been accused of taking lambs from farms, but it's likely that these were already dead, as boars tend to scavenge rather than actively hunt.

After a meal, a boar likes nothing better than a good wallow in the mud. This is done primarily to keep their skin free from parasites but, like domestic pigs, they seem actively to enjoy it.

Shout It Out Loud!

Boars are extremely communicative and vocal animals. They use a wide range of body movements, scents, grunts, cheeps and squeals to communicate with the rest of the herd, which is a matriarchal (female-led) group called a sounder. Males live alone, but during the mating season, from December to January, they become extremely active and noisy. Competitions between males contesting for the right to mate with females begin with a loud snorting challenge that can be heard from 18m (60ft) away. This may progress to a duel, in which the boar's tusks (in fact,

Wallowing in mud has several benefits: not only does it keep the boar cool, it also prevents infection and deters flies.

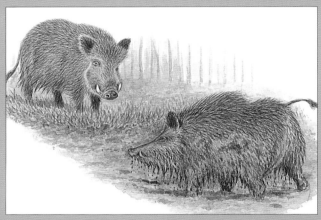

Another male attempts to muscle in on the wallowing hole. A fight seems inevitable.

The boars charge at each other, slashing and tearing with their tusks. These clashes often result in serious wounds.

Victorious, the defending boar chases away the injured and defeated intruder.

elongated lower canine teeth) play a decisive role. Duels can be violent and bloody affairs, and opponents are often badly injured, if not killed, by wounds inflicted by their opponent. Females (sows) will also attack other females within the sounder to establish dominance.

It's when they are disturbed that they are at their most dangerous, both to man and other animals. This is especially true of sows protecting their young, since they make dedicated mothers. Most boars will run away from humans unless they're cornered, but sows will charge any would-be aggressor to chase them away from the nest. Piglets stay within the sounder for about two years, when the young males leave once their tusks have developed.

Old Meets New

Wild Boar have been hunted for their meat for thousands of years. Today, boar hunts are still a common occurrence throughout Russia and the USA. The continuing popularity of boar hunting is mainly due to the boar's renowned aggressive nature, which offers a challenging hunt. Modern huntsmen generally use guns, but in India they still kill boars from horseback using long spears, called pig-sticks. This tradition harks back to the Middle Ages, when huntsmen on horseback would track down a boar, using large hunting dogs to wear down the boar. Alain of Lille, writing in 1202, describes how this was often an uneven contest for the dogs: 'There the wild boar, by its

Wild Boar habitats

murderous weapon of a tusk, sold its death to the dogs for many an injury.' Hunters would then finish the kill on foot, using long, barbed spears. The barbs were necessary to prevent the huntsman from being injured by the boar. Even if it were stabbed through the heart, the boar would often be so enraged that it would continue running up the spear in an attempt to gore the hunter with its tusks. Dead or alive, it seems that Wild Boars are just as dangerous!

Comparisons

The African Wart Hog is a close relative of the European Wild Boar. Much smaller in size, the Wart Hog also has less fur, which is the result of living in a considerably warmer climate. The most noticeable difference between the hog and boar are the hog's enormous tusks.

These are, in fact, enlarged canine teeth, which in most predators are the teeth that are used to deliver the killing blow. As such large curved 'teeth' are useless for this sort of job, their main purpose now is in mating displays.

Wart Hog

Wild Boar

ASIA

SOUTH
CHINA SEA

PACIFIC
OCEAN

ARAFURA
SEA

TIMOR SEA

CORAL
SEA

INDIAN
OCEAN

AUSTRALIA

SOUTHERN
PACIFIC
OCEAN

SOUTH
AUSTRALIAN
BASIN

TASMAN
SEA

SOUTHERN
OCEAN

The World's Oceans

The oceans are one of the world's last true wild frontiers, a huge and largely unexplored world full of fabulous — and often very fearsome – creatures.

~

Around 70 per cent of our planet's surface is covered by ocean – from the ice-choked Arctic to the tepid waters of the tropics. This vast body of liquid actually comprises one huge 'global' ocean but, over the centuries, cartographers (map-makers) and geographers have divided this colossal watery mass into numerous liquid 'continents'.

The largest of these is the Pacific Ocean, which covers a third of the world's entire surface. Next is the Atlantic, which is sandwiched between Europe and Africa in the east and the Americas in the west. At just half the size of the Pacific, the Indian Ocean is the last of the 'big three'. The warmest of all the world's waterways, the Indian Ocean touches the coastlines of Africa, India and China, and embraces the Red Sea and the Persian Gulf on its journey around the globe. The smallest of these marine environments are the Arctic Ocean, in the northern hemisphere, and the Antarctic Ocean in the southern hemisphere, although some geographers include these freezing waters as part of the other three.

These great masses of water are, in many ways, the liquid equivalents of the jungle, plains and tundra of the great landmasses. Each ocean has its own giant herbivores and cunning predators, its own fighters, bluffers and pretenders. Yet life in the dark depths of the oceans has also found its own unique solutions to the problems of survival, making this watery world a strange, surprising and often extremely dangerous place.

Blue-Ringed Octopus

The last thing a victim of this Pacific predator usually sees, before blindness and paralysis set in, are rows of pulsating blue rings. A Blue-Ringed Octopus may be no bigger than a hen's egg, but it carries one of the deadliest toxins known to science. A single adult possesses enough venom to kill ten humans, and there's no known antidote.

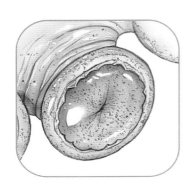

Suckers

On average, a Blue-Ringed Octopus has 40 suckers on each tentacle, which it uses to hold and grip prey, and to move itself along the ocean floor.

Key Facts

ORDER *Octopoda* / FAMILY *Octopodidae* / GENUS & SPECIES *Hapalochlaena maculosa*

Weight	90g (3oz)
Length	
Body	7.5cm (3in)
Tentacles	11.5cm (4^1/2in)
Sexual maturity	2 years
Mating season	Usually spring
Number of Eggs	Variable: usually a few dozen
Breeding Interval	Male dies soon after mating; female dies about a month after producing her eggs
Typical Diet	Crabs and other crustaceans
Lifespan	Probably 2 years

Mouth
Inside the octopus' sharp and powerful beaklike mouth is a rough, barbed tongue (called the radula), which is used to scrape and shred food.

Coloration
In the animal world, intense colours often warn of danger. The rings of the Blue-Ringed Octopus become prominent only when it is alarmed.

Comparisons

Octopus, squid and cuttlefish all belong to the animal Class *Cephalopoda* and, as would be expected, share many features. However, in addition to the 8 'usual' arms, squid and cuttlefish also have 2 additional tentacles, which can shoot out at lightening speed to capture prey. The Nautilus, which has a hard, spiral outer shell, also has tentacles – around 90 in total!

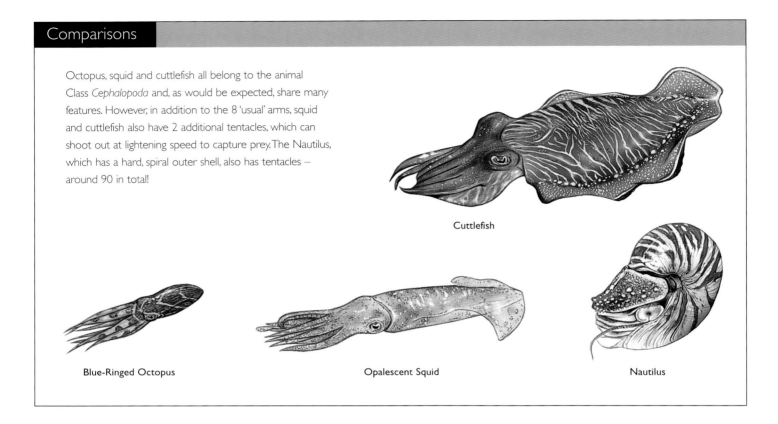

Cuttlefish

Blue-Ringed Octopus

Opalescent Squid

Nautilus

Blue-Ringed Octopuses are typically found in the waters off the coast of Australia, the Philippines and Indonesia. No one knows for certain just how many species there are, and so far only three have been verified: the Greater Blue-Ringed Octopus, the Lesser Blue-Ringed Octopus and the Blue-Lined Octopus.

Alien Visitors

Of all the strange animals that inhabit the oceans, the octopus seems the most other-worldly. There are an estimated 100 species of this remarkable creature, most of which make their homes in the sand and mud on the ocean bed. Octopuses belong to a group of animals called molluscs. Unlike oysters, which are also molluscs, octopuses have no hard outer shell. Indeed, their bodies are soft and boneless, so the octopus's brain and internal organs have to be protected by a sheet of muscular skin called a mantle.

At the end of the mantle, eight tentacles, covered in powerful suckers, extend outwards. These arms are extremely dextrous and are used not only to catch prey and to propel the octopus across the ocean bed, but also to perform complex tasks. In tests, octopuses have been proved to be extremely intelligent and able to utilize their tentacles with skill and ingenuity.

It is believed that some species of octopus – like the great giants of sea-faring legend – may grow to immense lengths, but the largest recorded is the Giant Octopus, which can reach 6m (almost 20ft) from tentacle to tentacle. The Lesser Blue-Ringed Octopus is one of the smallest, but don't be fooled: this tiny, tentacled terror still packs a mighty punch.

Blue for Danger

Every year, dozens of people around the coasts of Australia are poisoned by the Lesser Blue-Ringed Octopus. This small, grey-brown, water-bound killer may look harmless, but just placing a foot in water recently inhabited by one can cause numbness and tingling. And a bite from this octopus can cause death in minutes.

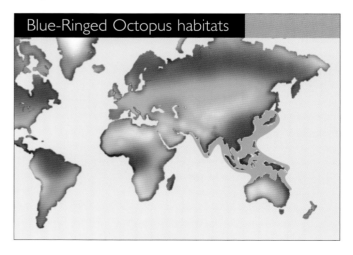

Blue-Ringed Octopus habitats

Like all octopuses, the Blue-Ringed Octopus has a sharp beak at the tip of its head. Venom is produced in glands, which sit just behind the brain. One of these glands carries toxins that seem intended specifically for use against prey, and the other – the man-killer – may be used in defence against predators. The problem for humans who are bitten is that, unless they see the octopus's warning blue rings, which pulsate in alarm when predators approach, they often literally don't know what's hit them. The bite wound can be so small that most people feel nothing until the toxin, called tetrodotoxin, starts to react. Symptoms start with dizziness and nausea. After a few minutes, the victim is paralysed and unable to breathe. The only way to survive a blue-ringed bite at this stage is for artificial respiration to be performed, until the poison works its way out of the body. This may take several hours.

A Deadly Embrace

At home in the ocean the Blue-Ringed Octopus is a fast and agile hunter, whose fatal toxin is just as effective against prey as it is against humans.

Its favourite meal is crab, and the octopus is quick to pounce when one strays too close. Wrapping up the victim in its powerful tentacles to hold it still, the octopus uses its sharp beak to bite through the crab's outer shell, and then injects toxin straight into the body cavity. In just a matter of seconds, the toxin paralyses the victim. This keeps the octopus safe from any potential injury that could be inflicted by the crab's only defence – its claws. The octopus now quickly devours its prey. Using its beak to break open the crab's outer carapace, it sucks out the crab's soft flesh and leaves only the empty shell behind. It's an effective, if grisly, technique.

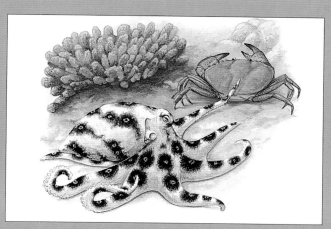

The octopus determines the size of a passing shore crab by touch, using a tentacle to gently probe the potential victim.

When it is satisfied that the crab can be overpowered, the octopus strike out, grasping its prey by the shell.

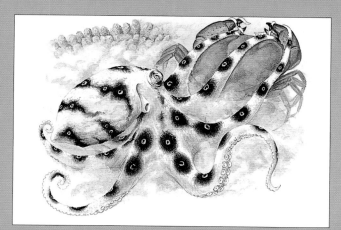

Quickly binding the crab claws to prevent a counterattack, the octopus draws its victim close to deliver the killing blow.

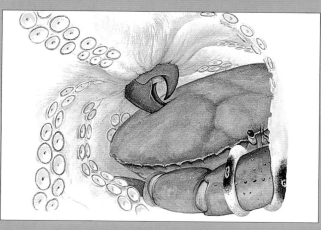

The beak of the octopus is powerful enough to pierce the crab's shell. It then injects saliva that kills the prey and liquefies its flesh.

Great White Shark

The Great White Shark is among the ocean's greatest predators. Sleek, beautiful and deadly, this magnificent hunter has captured the imagination of storytellers for centuries.

Teeth

Great White Sharks have two rows of teeth. New teeth can be grown to replace old, worn-out teeth every couple of weeks, if needed.

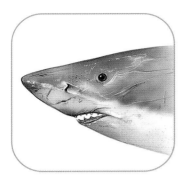

A bigger bite

As a shark moves in for the kill, it lifts its snout up. This pushes its lower jaw into alignment with the upper, which results in a bigger bite capacity.

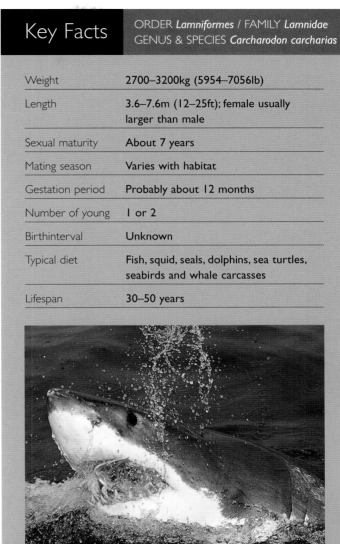

Key Facts	ORDER *Lamniformes* / FAMILY *Lamnidae* GENUS & SPECIES *Carcharodon carcharias*
Weight	2700–3200kg (5954–7056lb)
Length	3.6–7.6m (12–25ft); female usually larger than male
Sexual maturity	About 7 years
Mating season	Varies with habitat
Gestation period	Probably about 12 months
Number of young	1 or 2
Birthinterval	Unknown
Typical diet	Fish, squid, seals, dolphins, sea turtles, seabirds and whale carcasses
Lifespan	30–50 years

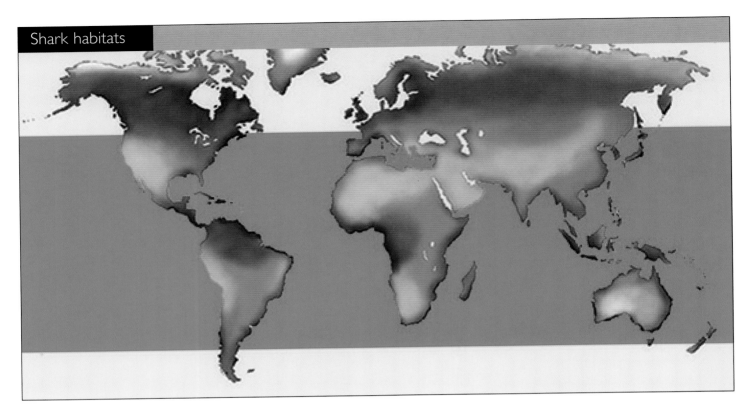

Shark habitats

Great Whites are found throughout warm, temperate waters, particularly in the deep oceans or around the coastlines off Australia and northeastern America. Since seals are among their favourite food, they also often gather near seal breeding grounds, such as Shark Alley, which is between Dyer Island and Gyser Rock, just off the coast of South Africa's Western Cape province. Here Great Whites have been seen attacking not just seals, but leaping completely out of the water to catch sea birds in an incredible display of marine acrobatics.

A Killing Machine
The Great White is designed for both speed and attack. Its grey-topped body is shaped like a torpedo, while a crescent-like tail helps to propel it through the water at incredible speeds. The first dorsal fin, which is the one often seen above the waterline, and pelvic and pectoral fins help keep it stable and manoeuvrable. Its teeth are triangular, with serrated edges, like a saw, for tearing through flesh and bone. These offensive weapons are so important to a shark that, if one is broken, a replacement can be grown within a matter of days. Like all sharks, the Great White has excellent senses and it is believed that they can actually 'smell' the electrical fields given off by other fish. Using specially adapted pores in their head (called Ampullae of Lorenzini), which are sensitive to electric fields, the Great White can locate its prey in even the darkest waters.

Such a specialist design does, however, have its flaws. Most fish breathe by pumping oxygen-rich water over their gills. Some sharks can't do this, and have to force water over their gills by constantly moving through the water. If they stop moving, they'll drown!

Fighting for Survival
For the Great White, as for other deep-sea sharks, the fight for survival may begin even before birth. During mating, the male shark uses two organs called 'claspers', to release

Comparisons

Dwarfing the great white in size is the massive – and aptly named – Whale Shark. Although it looks fearsome, this member of the Lamnidae Family is truly a gentle giant. Like many true whales, it feeds on plankton. These are tiny plantlike organisms, which it eats by straining through its platelike teeth.

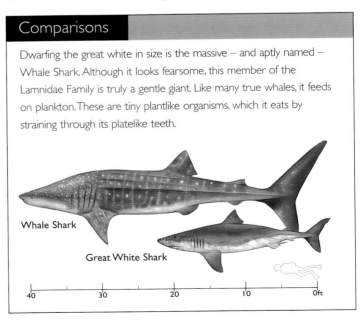

Whale Shark

Great White Shark

40   30   20   10   0ft

sperm into the female to fertilize her eggs. These eggs hatch inside the female, and the immature sharks immediately begin to fight, and even eat each other. Little is known about the early years of most species of shark, but this practice has been recorded in both tiger and blue sharks and is likely to be true for all sharks that give birth to live young. This grim practice is called uterine cannibalism. When the surviving pups are born, in litters of between seven and nine, they may measure only about 1.2 to 1.5m (4–5ft), but they're already practised killers.

Hunter and Hunted

Sharks and rays are part of an ancient group of fishes that have lived in the earth's oceans for almost 400 million years. During this time, species such as the Great White Shark have become such efficient predators that they now sit at the very top of the ocean's food chain. In water, the only animal that can compete for food directly with the great predatory sharks is the giant Orca, or Killer Whale.

With such huge jaws, just one bite from a Great White can cause a wound 28cm by 33cm (11–13in), which is easily enough to kill. No surprise, then, that it has gained such a fearsome reputation. Yet, according to surveys, the Great White has been responsible for only 61 human deaths in the past 100 years, which makes it less dangerous than the domestic dog. During most recorded attacks on people by Great Whites, the shark seems inquisitive rather than aggressive, often swimming away after it has tasted a 'sample'. In fact, the Great White would seem to have more to fear from humans than we have from them, since it's now on the endangered species list in South Africa, and protected along the Florida coastline.

Using its impressive array of senses, the Great White Shark detects a nearby dolphin swimming through the ocean.

Following the traces of scent in the water, the shark attacks with exceptional speed. At the last moment, the mouth opens and the eyes roll back.

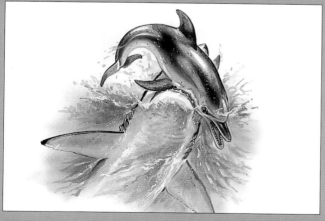

The jaws close around the victim as the shark thrashes its tail. This drives the head from side to side, sawing the sharp teeth through flesh.

The shark withdraws a short distance and waits for its victim to die of shock and blood loss. It will soon return to the carcass to devour more.

Orca

Orcas are known as the 'wolves of the sea' and, just like their land-bound namesakes, Killer Whales have learnt that hunting in packs can be a very effective – and deadly – tactic. Even Blue Whales, which are the largest animals ever to have lived, are regular additions to an Orca menu.

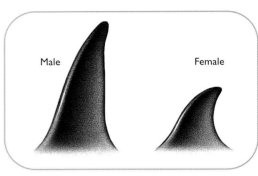

Dorsal Fin
To tell a male from a female Orca, simply look at the dorsal fin. A female's fin is about half the size of a male's.

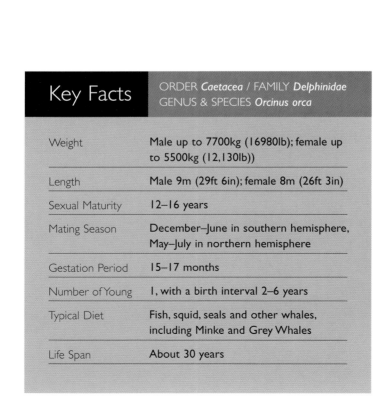

Key Facts	ORDER *Caetacea* / FAMILY *Delphinidae* GENUS & SPECIES *Orcinus orca*
Weight	Male up to 7700kg (16980lb); female up to 5500kg (12,130lb))
Length	Male 9m (29ft 6in); female 8m (26ft 3in)
Sexual Maturity	12–16 years
Mating Season	December–June in southern hemisphere, May–July in northern hemisphere
Gestation Period	15–17 months
Number of Young	1, with a birth interval 2–6 years
Typical Diet	Fish, squid, seals and other whales, including Minke and Grey Whales
Life Span	About 30 years

Athleticism
Orcas perform remarkable feats of athleticism when hunting. Some regularly throw themselves out of the water to grab prey.

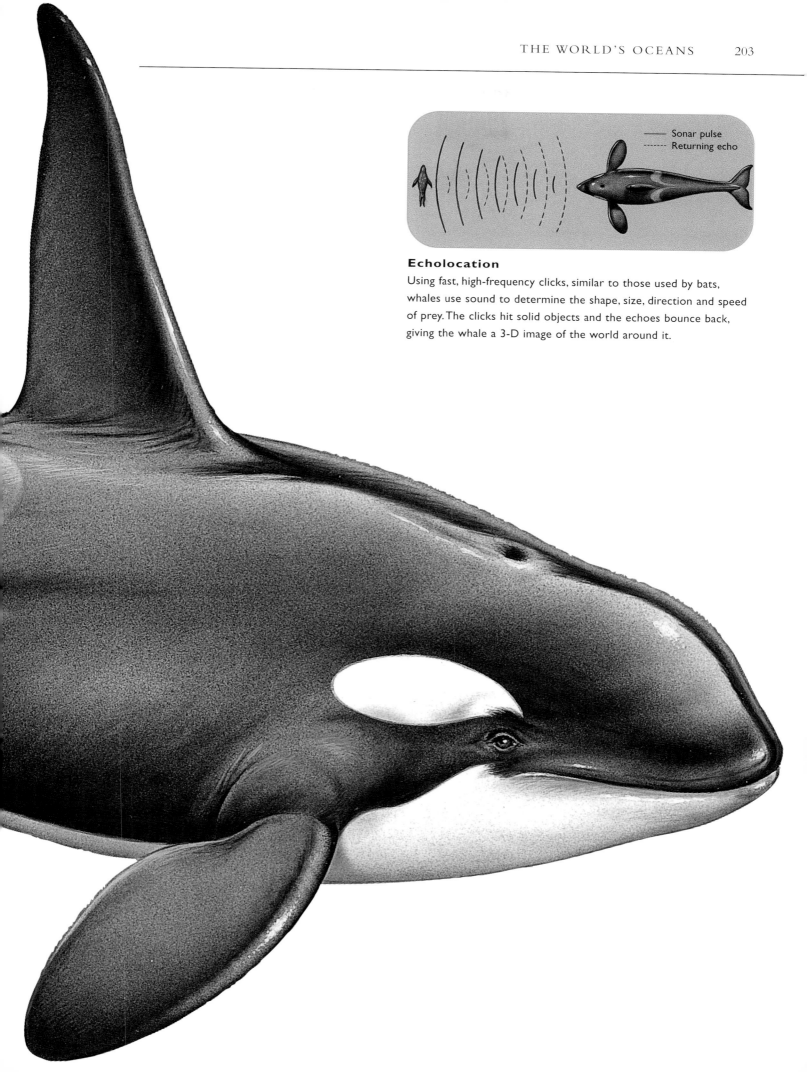

Echolocation

Using fast, high-frequency clicks, similar to those used by bats, whales use sound to determine the shape, size, direction and speed of prey. The clicks hit solid objects and the echoes bounce back, giving the whale a 3-D image of the world around it.

Comparisons

Unlike the giant Orca, Minke Whales do not have teeth. In common with many true whales, they feed on plankton, which are tiny plantlike organisms. The Minke is able to do this because it has a specially adapted fibrous plates in its jaw, called a baleen. It uses these to 'catch' the plankton.

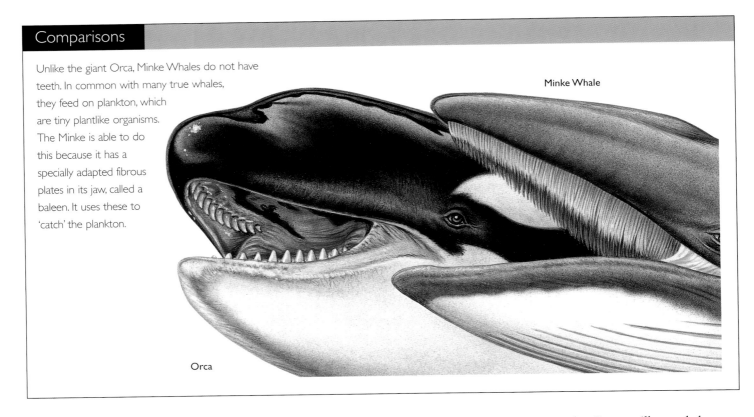

Minke Whale

Orca

Despite their name, Killer Whales are actually the largest member of the dolphin family. Found throughout the world's oceans, Orcas are equally at home in the Arctic or the tropics, and may cover many thousands of kilometres in search of seasonal prey.

Power to the Pod!

Orcas are social animals and live in groups called pods, which will include between five and thirty other whales. Occasionally, these pods may form even larger groups called aggregates, which seem to gather mainly for hunting or breeding purposes.

Social organization within the Orca pod is based on a hierarchy of dominant males and females, but all members will work together to protect the young or sick. Killer Whale society is highly structured and complex, and they are among the few marine animals who have been seen actively to teach and discipline their young.

Sophisticated Hunters

There's strength in numbers, and few marine animals can withstand an attack from an Orca pod. Great White Sharks, penguins, Sea Lions and walruses are all part of the diet. Polar Bears, moose and even other Killer Whales have also been found in Orcas' stomachs, although herring comprise more everyday fare.

During the hunt, an Orca pod will work together with clockwork efficiency, each member taking on a different role. When fish are the target, the Orcas will corral the shoal into a tight ball, displaying their white bellies, to startle the fish and force them towards the water's surface. Staying below the shoal, they will then stun the fish with their huge, falcate (sickle-shaped) tails and then pick them off, one by one.

With larger prey, more sophisticated tactics may be used. An Orca needs to eat around 5 per cent of its body weight every day, and will perform remarkable feats of athleticism to take larger prey. Some, for example, regularly throw themselves out of the water (a technique called breaching) to grab penguins or seals straight from the ice.

Orca habitats

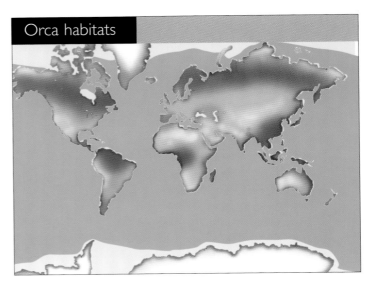

Spying a seal dozing on an ice floe, the three Orcas close in for a better look...

Working as a team, two of the Orcas swim to the floe and suddenly tip it towards the other side.

The startled seal slides off the floe and directly into the jaws of the waiting Orca. The seal has no chance of escape against such a well organized attack.

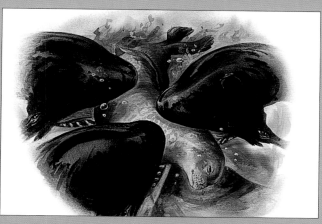

The other two Orcas quickly join in the feast of plump seal, tearing it to shreads. The teamwork has paid off.

With dangerous prey, such as walruses, a more cautious approach is required. The pod will attempt to isolate a family group, and then panic the adults into diving so that vulnerable young are left alone. The young walrus will then be stunned by a blow from the tail and torn into hunks by the Orca's huge (7.6cm/3in) interlocking teeth. Orcas can't chew, but their mouths are large enough to swallow a seal whole.

Super Powers

Orcas are natural athletes, whose bodies are perfectly adapted for a life in the oceans. They're fast and agile swimmers. Generally a Killer Whale can cruise at about 3–9km/h (2–5mph), but they've been clocked at 56km/h (35mph), which makes them the fastest swimming marine mammal. As they have lungs, not gills, Orcas need to come to the water's surface regularly to breathe. Yet they can hold their breath for up to ten minutes when diving, by slowing down their heart rate to save oxygen. When resting, another remarkable adaptation kicks in. If an Orca were to fall asleep, it would drown. So Killer Whales never sleep, they just rest one half of their brain at a time!

Orcas are extremely vocal animals, and communicate continually with the rest of the pod through a series of whistles and grunts. Using extremely fast, high-frequency clicks, similar to those used by bats, whales can also use sound to identify their prey with incredible accuracy, determining not just shape and size, but also direction and speed. The clicks hit solid objects and the echoes bounce back, giving the whale a 3-D image of the world around it. This 'echo-location' is just another of the whale's seemingly 'super' abilities.

Portuguese Man-of-War

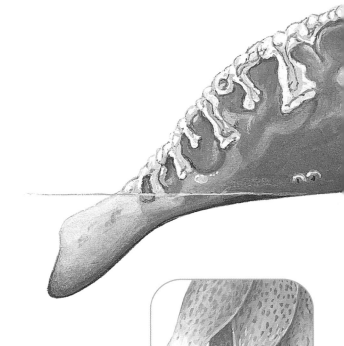

Nematocysts

Cells within the Portuguese Man-of-War's tentacles discharge barbed stingers. These contain a powerful toxin, which can be fatal to any animal or human unlucky enough to brush past them in the water.

Digesting polyps

Once their prey has been disabled by poison, it is slowly digested by organisms specially adapted for the job.

Key Facts

ORDER *Hydrozoa* / FAMILY *Siphonophora* GENUS & SPECIES *Physalia physalis*

Weight	Variable; mass made up mainly of water
Float length	Up to 30cm (12in)
Tentacle length	Usually up to 10m (33ft) but may reach 60m (197ft)
Sexual maturity	Unknown
Breeding season	Year-round
Number of young	Possibly millions in a single lifetime
Birth interval	Releases medusae (sexually active stages) almost continuously
Typical diet	Small fish and fish larvae, crustaceans, plankton
Lifespan	Unknown

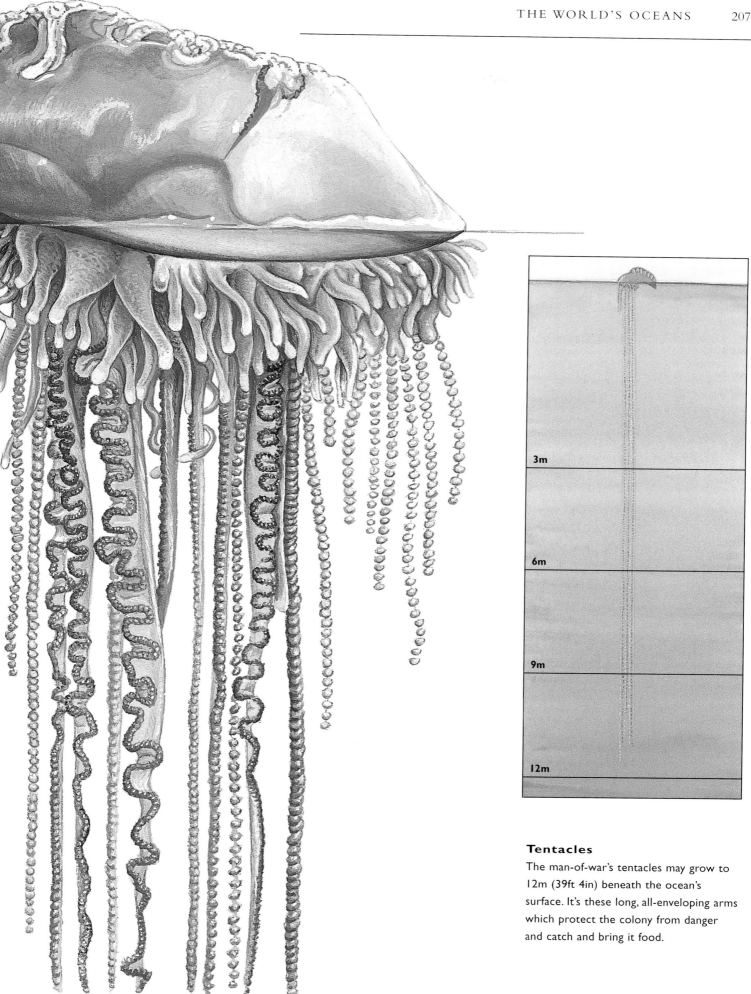

Tentacles

The man-of-war's tentacles may grow to 12m (39ft 4in) beneath the ocean's surface. It's these long, all-enveloping arms which protect the colony from danger and catch and bring it food.

3m

6m

9m

12m

Portuguese Man-of-War habitats

The Portuguese Man-of-War is often found in warm waters around Australia and Hawaii. It is also frequently washed close to shore during storms, which is when it presents the greatest threat to humans.

A Life on the Ocean Waves

Wherever the wind blows, the Portuguese Man-of-War follows. This complex creature was named after an early type of warship and, like a ship, it's carried across the world's oceans by the wind. The man-of-war floats using a large, crested gas bag, which acts like a sail above the waves. This translucent (see-through) structure may extend as much as 15cm (6in) above the water, but can, just like a sail, be deflated during storms to prevent the man-of-war from being damaged.

The man-of-war may look like one individual, but this ocean nomad is actually a colony of separate, but interdependent, life forms. The floating canopy, called the pneumatophore, is the original member of the colony. This one individual reproduces all the other additional colony members. As if that wasn't startling enough, each new colony member has a distinct role to play in survival, reproduction, hunting and feeding.

All Hands on Deck!

If we imagine the man-of-war as a real mini ship, then the marines and catering corps would be the dactylozooids. These are the tentacles that hang beneath the floating platform. Typically a man-of-war's body may be no more than 8–30cm wide (almost 3–12in), but its tentacles may grow to 12m (39ft). It's these long, all-enveloping arms that fight for the colony, protecting the whole from danger and catching and bringing it food. Cells within these tentacles discharge barbed stingers into enemies and prey alike. These stings contain a powerful toxin and can be fatal to any human unlucky enough to brush past them in the water. Even a dead man-of-war, washed up on the beach, is a serious health threat and should be avoided.

For the most part, though, it's fish that are on the receiving end of this deadly package. All, that is, except

Comparisons

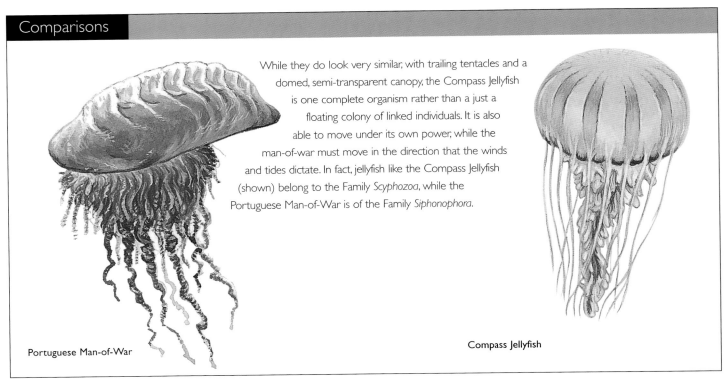

While they do look very similar, with trailing tentacles and a domed, semi-transparent canopy, the Compass Jellyfish is one complete organism rather than a just a floating colony of linked individuals. It is also able to move under its own power, while the man-of-war must move in the direction that the winds and tides dictate. In fact, jellyfish like the Compass Jellyfish (shown) belong to the Family *Scyphozoa*, while the Portuguese Man-of-War is of the Family *Siphonophora*.

Portuguese Man-of-War

Compass Jellyfish

some species of jack fish, which seem to be immune. The jack fish has developed a symbiotic partnership with the man-of-war: a 'you scratch my back, I'll scratch yours' relationship that benefits both parties. The jack fish actually attracts larger prey into the tentacles and is, in turn, rewarded by a share of the spoils! Once the toxin has done its work, the tentacles take the food to the gastrozooids. These are the man-of-war's engines. They digest the dead body, using a range of proteins called enzymes to break down the flesh.

Be My Buddy

The gonozooids are the ship's workshop, taking care of reproduction and 'building' more men-of-war. This amazing process is made possible by a form of asexual reproduction called budding.

Most animals reproduce when a male and female come together to mate. New life is created by combining the male's sperm and the female's egg. It was long believed that the man-of-war was a jellyfish, but it's actually a member of the phylum *Cnidaria*, a group of marine organisms that includes sea anemones and sea firs. Animals such as this reproduce when a small part of the individual breaks off and develops as an independent life form. The new individual is genetically identical to the original parent. Such a process has its disadvantages. When a single individual is replicated, so too are any deficiencies, whereas animals born by sexual reproduction inherit the genes of both parents. However, it is a fast and efficient method of reproduction and men-of-war can multiply so quickly that swarms are often seen in the warm waters of the Gulf Stream.

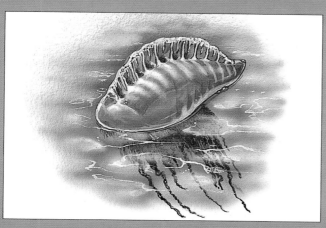

The Portugese Man-of-War is the nomad of the Atlantic, swept along by the currents of the ocean. Likewise, the process of hunting involves no effort.

By chance, a small fish brushes against the long tentacles. The man-of-war reacts instantly, stunning the fish with a dose of poison.

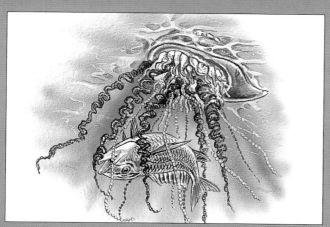

The tentacles contract, pulling the helpless fish upwards towards the thicker, shorter feeding polyps.

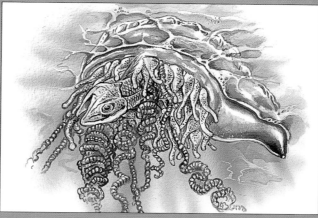

The process of digestion begins. The feeding polyps break down the victim's flesh and distribute the nutrients through the colony.

Puffer Fish

With its elongated body and large head, the Puffer Fish may look curiously comic, but this slow and ungainly animal carries a poison 1200 times more toxic than cyanide. Any predator that takes a step too close is dicing with death.

Key Facts	ORDER *Tetradontiformes* / FAMILY *Tetradontidae* GENUS & SPECIES *Various*	
Weight	Up to 6.5kg (14lb 5oz)	
Length	8–60cm (3–24in); exceptionally 90cm (35in)	
Sexual maturity	Varies according to species	
Spawning season	Varies according to region and species	
Number of eggs	200–300	
Breeding interval	Several spawnings a year	
Typical diet	Corals, crustaceans and molluscs; occasionally fish	
Lifespan	Not known	

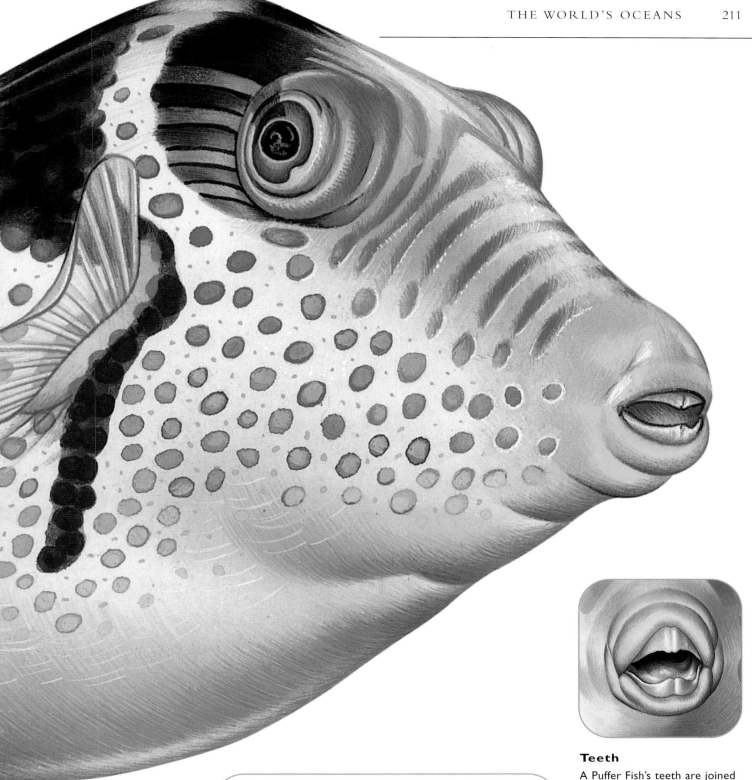

Teeth
A Puffer Fish's teeth are joined together to form a tough, beaklike opening. This is used to crack open the shells of crabs and molluscs.

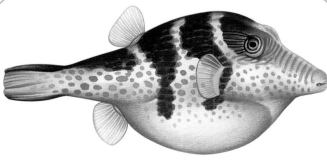

Inflated body
Some species of Puffer Fish reach 60cm (24in) in length, but by filling their stomach with water or air, puffers can 'grow' 300 times larger.

Puffer Fish habitats

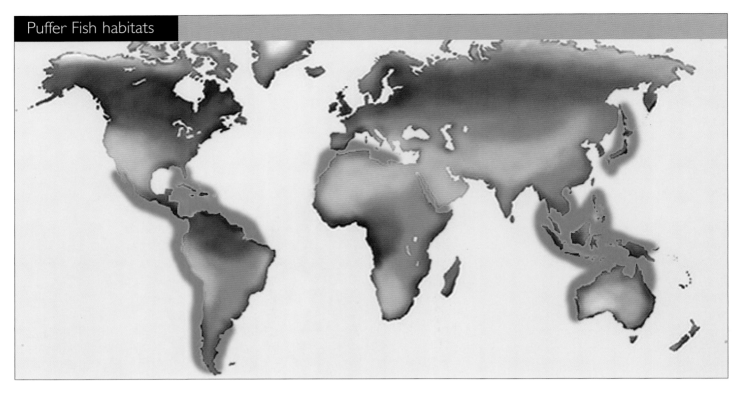

There are an estimated 120 species of Puffer Fish. These live mainly in tropical waters, although a few freshwater species are found in Africa and Asia.

Dangerous Dining

Japanese diners are renowned for their sophisticated appreciation of food, but one popular delicacy – fugu – appeals only to the most adventurous.

Fugu is the Japanese name for the Puffer Fish, and a meal of fugu is an occasion, surrounded by ceremony. Before the diners eat, the chef presents them with the fish for their approval. He or she then returns to the kitchens to prepare the fugu. First, the fins are served in hot saki (rice wine). Then the skin is removed and prepared with a salad. Finally the fish is filleted (the bones are removed) and wafer-thin, raw slices are served to the diners. Fugu has a delicate flavour, and if prepared correctly, the natural toxins in the fish's body will give the diner a warm, tingling sensation that most people describe as pleasant. If prepared incorrectly, however, the results can be fatal. All fugu chefs have to be specially trained and licensed following a written and practical test. Even after the chef

Two male Puffer Fish face each other off over territory on the rocky seabed.

Both fish erect a thin row of spines on their backs and a ridge on their bellys.

has qualified, preparation requires a meticulous thirty-step process. Despite these stringent rules, up to a hundred people every year are poisoned by eating badly prepared puffer fish. Perhaps it's not surprising then that the word fugu, as well as meaning 'to swell', means good luck!

Comparisons

Puffers are not the only fish to use such an ingenious defensive technique. Porcupine Fish, which are found in warm Tropical waters, especially around coral reefs, can also inflate their bodies to deter potential aggressors.

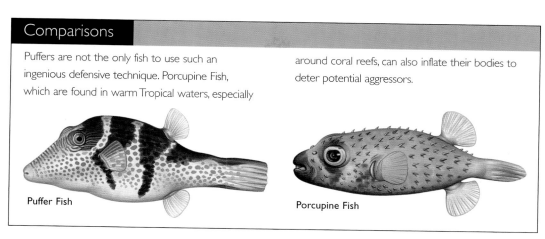

Puffer Fish

Porcupine Fish

Deadly Defences

In the wild, the Puffer Fish's toxins form part of its defence mechanism. The fish's entire body, especially its liver, is seeped in tetrodotoxin, the same type of poison used by the Blue-Ringed Octopus. Effects can be felt within 20 minutes, and can be fatal in less than six hours. Symptoms vary but begin with dizziness and nausea and, in extreme cases, the victim may become paralysed. Tetrodotoxin victims are often unable to move or respond to stimuli, but later report that they could still see and hear everything quite clearly. In Haiti, 'witch doctors' are believed to use a powder made from the spines of Puffer Fish to create so-called zombies – victims who appear to be dead, but later miraculously come back to life.

The Puffer Fish's most obvious and reliable defence, though, depends on its ability to 'puff up' or inflate its body. Some species are 60cm (24in) long, but by filling its stomach with water or air a puffer can make itself appear much larger. The underside of a puffer's body is covered in poison-laden spines. Once inflated, these stick out, making a formidable mouthful for any would-be predator.

Unsociable Habits

Puffers vary in size, shape and coloration between species, from the tiny Valentines Puffer Fish, to the freshwater Giant Puffer Fish of Africa. A young puffer will usually spend most of its time in shallow waters, where they're relatively safe, but once it has grown, the puffer's predatory instincts take over. They might be better known for their defensive capabilities, but most species of puffer fish are notoriously aggressive.

Puffers have a robust beaklike jaw and extremely sharp teeth. These are used principally for tearing at coral, or for crushing the shells of clams and crabs, which are the puffer's natural prey. Yet it may also be put to more surprising use: so powerful is this beak that there are stories of puffers inflating themselves after being a swallowed by a predator and, once the predator has died, eating their way out of its stomach!

Puffers are popular pets, but are difficult to keep in captivity, as they will attack any other fish placed in the aquarium with them.

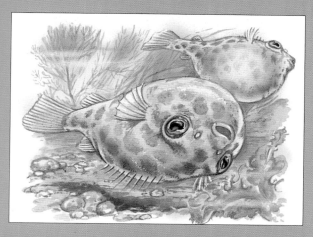

To look as large and intimidating as possible, each fish inflates their bodies and swims upside-down.

Whoever puts on the most impressive performance usually wins, forcing the loser to withdraw.

Sea Snake

Amongst all the strange and beautiful life forms that make their homes in the ocean's own version of forest — the coral reefs — is one unexpected visitor. Like its land-bound relatives this sinewy serpent is a proficient killer, ready to use its own lethal load of venom on any unsuspecting prey.

Key Facts

ORDER *Squimata* / FAMILY *Elapidae*
GENUS & SPECIES *Hydrophiinae, Laticaudinae*

Weight	1.5–2kg (3lb 5oz–4lb 6oz)
Length	Up to 2m (6ft 6in)
Sexual maturity	2–3 years
Breeding season	All year
Gestation period	150–180 days (*Hydrophiinae* only)
Number of young	2 to 12 (*Hydrophiinae* only)
Number of eggs	5 to 10 (*Laticaudinae* only)
Incubation period	Not known
Birth interval	Not known
Typical diet	Mainly fish; some species eat fish eggs
Lifespan	Up to 20 years

Nostrils

Nostrils on top of the Sea Snake's head allow it to simply skim the water's surface when it needs to breathe. This is safer than raising its head completely out of water, which would leave its body vulnerable.

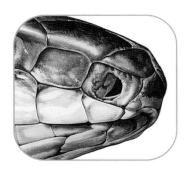

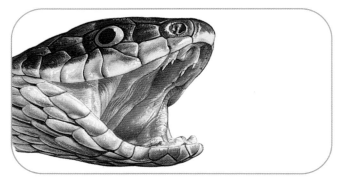

Fangs and Venom

The Sea Snake's relatively small fangs are used to deliver a dose of venom to its victim. Sea Snake poison is one of the most toxic, as the snake can't afford to let its prey crawl away and die somewhere out of reach.

Sea Snake habitats

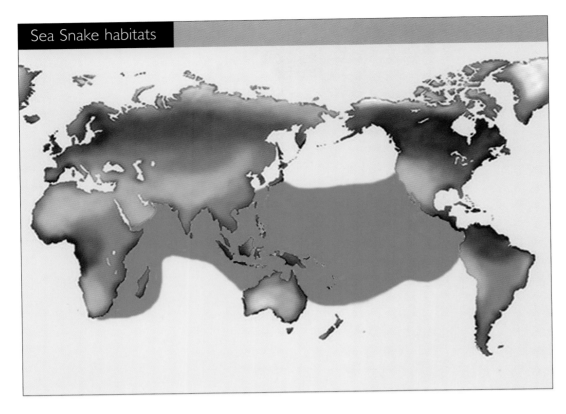

gills, the organs that a fish uses to breathe. As they swim, water passes into a fish's mouth and out through the gills, where specialized filaments extract oxygen from the water. Like land-dwellers, snakes have lungs, or rather one long lung, which stretches the length of their body and doubles as a swim bladder to keep the snake afloat. The reason that they're able to say submerged for such a long time is due to their relatively slow metabolism and their skin. In some species, this skin has been adapted to absorb oxygen from the water to extend their dive time.

There are about fifty species of Sea Snake, although zoologists are still divided as to whether Sea Kraits should be classified as Sea Snakes.

Just One Big Breath…

Perhaps the most surprising fact about the Sea Snake is that it's an air-breathing reptile. This amazing snake can be found throughout the tropical waters of the Indian and Pacific Oceans, although they settle on reefs close to the shore when they're not swimming. They can hunt submerged beneath the waves for an hour, diving to 90m (295ft) in pursuit of a meal. They may even sleep underwater. Yet this ocean-going reptile is not a natural marine inhabitant. The snake doesn't, for example, have

Extra Strong

One look at the bright patterning on a Sea Snake's body tells you instantly that this animals is dangerous. In fact, Sea Snake venom is one of the most toxic snake venoms known; and it needs to be. In the ocean, the snake can't wait for its prey to crawl away and die, since the body would be washed away on the current. Its poison needs to be potent enough to do the job immediately.

A Sea Snake's favourite food is fish and eels, and most 'true' Sea Snakes can often be found prowling the coral reefs where their victims hide out in holes and crevices. Luckily the snakes are not naturally aggressive but they can, and will, bite humans if provoked, and the bite is potent enough to be fatal.

Comparisons

Yellow-Lipped Sea Snake

Yellow-Bellied Sea Snake

Snakes like the Yellow-Bellied Sea Snake belong to the Family *Hydrophiinae*, whose members live their lives entirely at sea. Close relatives such as the Yellow-Lipped Sea Snake, which belong to the *Laticaudinae* Family, still spend some time on land. These 2 sub-species are so closely related that many zoologists are still divided as to the correct classification.

A Change for the Better?

It is believed that, in the distant past, the ancestors of today's land–dwellers lived in the oceans. The transition from the sea took many millennia, but it probably began simply enough. A group of fish began to hunt closer to the shore, where the pickings were easier. Over time, those fish that were better able to tolerate the shallows, became more successful. They had more food, so they produced more offspring, who in turn inherited these survival traits from their parents.

Today, there are many curious animals in the world that seem to have been caught in mid-transition. The walking catfish, for example, can live for days on dry land and is able to 'walk' (really a slow shuffle) from one water source to another. The flying squirrel is another animal that seems undecided about its true vocation. A natural land–dweller, it can nevertheless glide quite well from tree to tree, using folds of skin stretched between their arms and legs in the place of true wings. Similarly, the Sea Snake seems to be in the process of returning to the oceans from which its ancestors originally crawled. By flexing their muscles, land snakes produce a series of ripples that move them forwards. A Sea Snake's body is flatter, and has a paddle-like tail. This is ideal for swimming but makes the snake immobile on land. Yet some sub-species, such as the Banded Sea Krait, still need to come onto dry land to mate and lay eggs.

Before diving down to the coral to hunt, the snake takes a lungful of air.

Most of the fish that swim around the coral are small and slow, and therefore ideal prey for the Sea Snake.

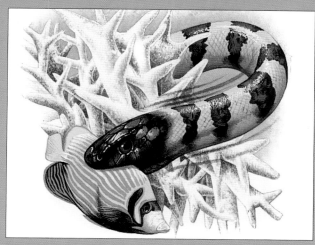

A passing Angel Fish becomes the snake's next victim. It is defenceless against the sharp fangs and lethal venom.

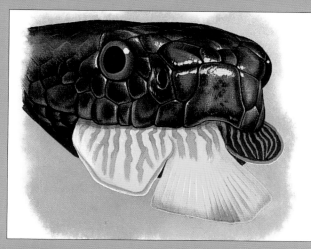

The helpless Angel Fish is quickly paralysed and promptly swallowed whole.

Stingray

If the Orca is the 'wolf of the seas',
and the Sea Snake its cobra, then the
Stingray is the marine equivalent of
the scorpion. When you live in the
ocean jungle, offence is often the best
form of defence, and for the Stingray
this means a tail packed with
pain-inducing poison.

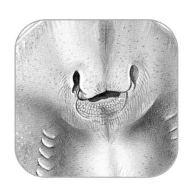

Mouth
Stingrays spend much of their
time combing the ocean floor
in search of a meal. Some
Stingrays have their mouth in
their stomach, and will swallow
clams and small crabs whole,
spitting out the shell later!

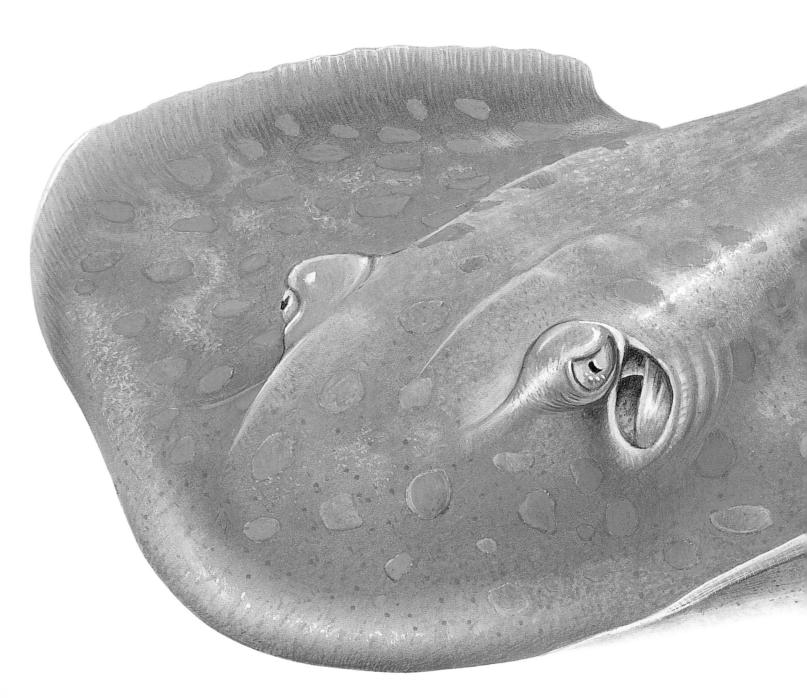

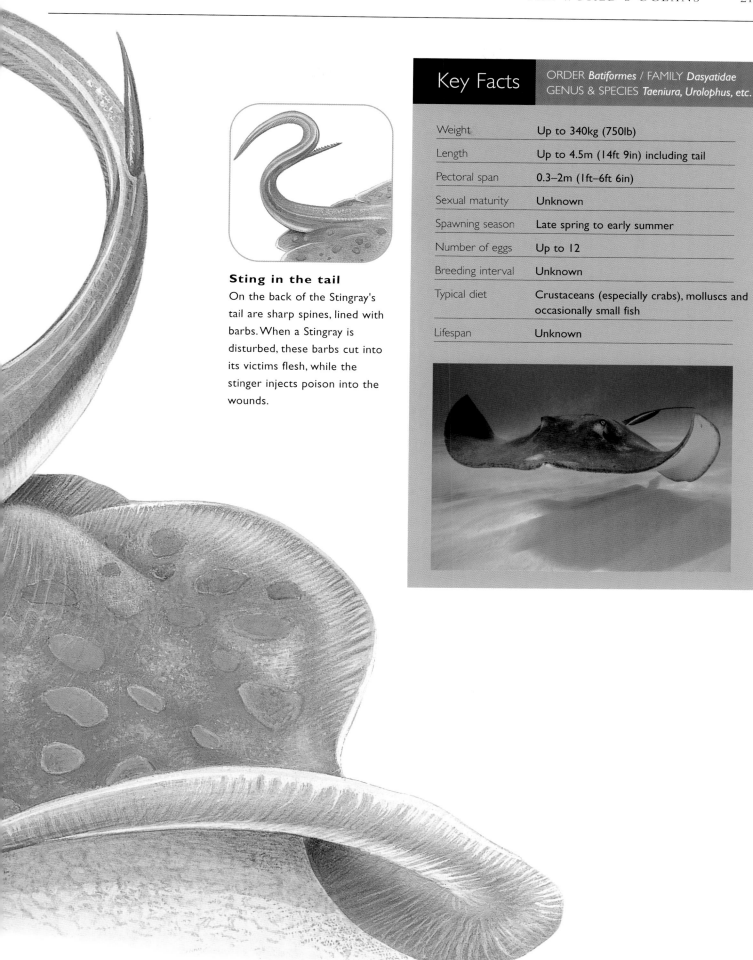

Key Facts	ORDER *Batiformes* / FAMILY *Dasyatidae* GENUS & SPECIES *Taeniura, Urolophus, etc.*
Weight	Up to 340kg (750lb)
Length	Up to 4.5m (14ft 9in) including tail
Pectoral span	0.3–2m (1ft–6ft 6in)
Sexual maturity	Unknown
Spawning season	Late spring to early summer
Number of eggs	Up to 12
Breeding interval	Unknown
Typical diet	Crustaceans (especially crabs), molluscs and occasionally small fish
Lifespan	Unknown

Sting in the tail

On the back of the Stingray's tail are sharp spines, lined with barbs. When a Stingray is disturbed, these barbs cut into its victims flesh, while the stinger injects poison into the wounds.

Comparisons

Stingrays generally have wide, flat, pancake-shaped bodies, and long, flexible tails. It's estimated that this simple and successful design has been adopted by around 100 species, most of which live in warm, tropical waters. Stingrays are bottom feeders and so generally come in shades of yellow and brown, which helps them to blend in with the ocean floor. All stingrays carry poison but some, like the spectacularly colourful Blue-Spotted Ray, choose to advertise the fact with easier-to-spot colouration.

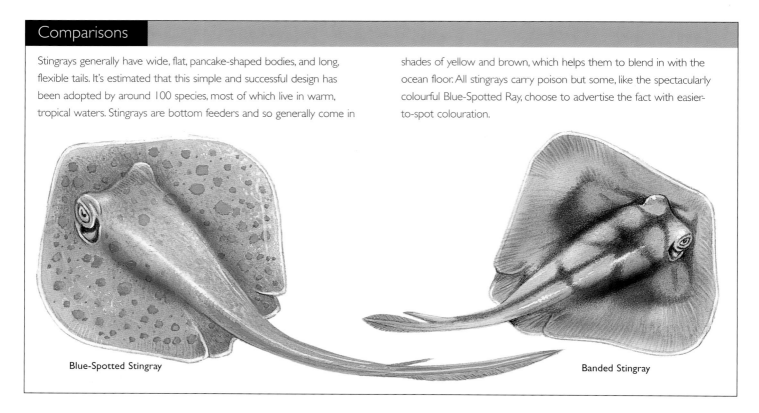

Blue-Spotted Stingray

Banded Stingray

Rays belong to the same class of animals as sharks and skates, and share many features, such as giving birth to live young. Like sharks and skates, rays are an extremely old and successful species, who have adapted so well to the world around them that, while most make their homes in warm waters, some, such as the Atlantic Stingray, can even tolerate freshwater estuaries and rivers.

What Shocking Relatives!

There may be as many as 600 species of rays, but not all are stingers. Some, such as the beautiful Manta Ray, which is the largest of all the rays, are relatively harmless. This graceful but gentle giant glides through the ocean on 4.5m (15ft) 'wings' – really elongated fins – and feeds on plankton and small invertebrates. Other rays, like the Lesser Electric Ray, are, as the name suggests, capable of producing a powerful electric shock. This is an effective way of warning away would-be predators. Yet this is not its primary use. Electric rays have small mouths, which makes it difficult for them to catch food, but by using specialized organs, situated just behind the head, an electric ray can generate an electric charge of between 14 and 37 volts. This is powerful enough to stun prey. In fact, they have been known to knock out a fully grown human.

Whip Lash

An estimated 100 species of rays are Stingrays, and these members of this ancient fish family have their own method of defence. Stingrays are flat fish, with squashed-looking bodies and long, tapering tails. On the tip of these tails may be one or more sharp spines, which are lined with thornlike barbs. At the base of these spines are poison glands. When a stingray is disturbed or threatened, they whip up their tail with incredible speed. The tail is thrust towards their victim and the barbs produce deep, jagged lacerations into which the poison is injected. Like scorpion venom, the ray's sting contains chemicals specially designed

Stingray habitats

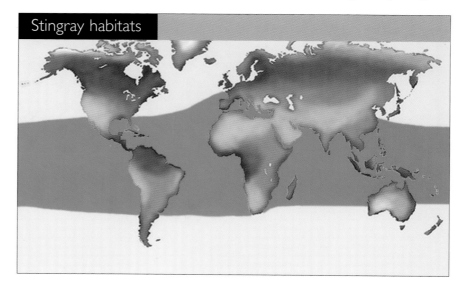

to cause pain and warn off any potential attacker. This toxin is extremely potent: if humans are unlucky enough to be stung, it can cause violent cramps, vomiting and, in rare instances, death. All of this proves, once again, that dangerous animals are not necessarily those armed with sharp teeth, claws or voracious appetites.

A Visit to the Beauty Parlour

Hunting is a leisurely affair for a Stingray. Most of these species are bottom feeders, who specialize in eating molluscs, crustaceans, worms and small fish from the ocean floor. The ray's sandy coloration provides great camouflage, and this patient predator is able to lie hidden in soft sediment until a meal comes its way. They conceal themselves by flicking sand or mud onto their body using their pectoral fins until only their eyes, water vents and tail are visible – and only then to very observant prey or predators. Some Stingrays actually have their mouth in their stomach, and will swallow clams and small crabs whole, spitting out the shell later!

However, not all small fish are on the menu. Stingrays will often wait in line while Spanish Hogfish and Bluehead Wrasse give their skins a clean. Such 'cleaning stations' are regularly visited by some of the ocean's most feared hunters, including Tiger Sharks and Great Whites. In eating ticks and mites, these small fish provide such an important service that all hostilities are suspended. These cleaner fish will even work around or in the mouths of some of the larger carnivores with no fear of being eaten. For both parties, it's a 'win–win' situation. The cleaners get a free meal and the rays not only rid themselves of irritating skin parasites, but get a relaxing massage too!

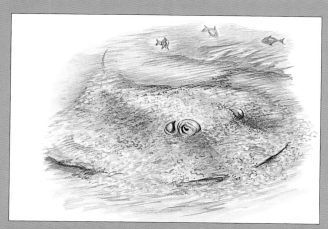

The Stingray has a natural ability to hide. Aside from its flat shape and camouflaged patterning, it also flicks sand over itself to blend in to the sea floor.

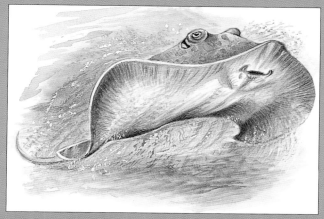

The Stingray propels itself by beating its pectorial fins, gracefully gliding through the water.

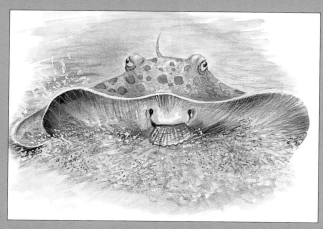

The Stingray gets at shellfish in the sand by beating its 'wings' and squirting water from its mouth to remove their cover.

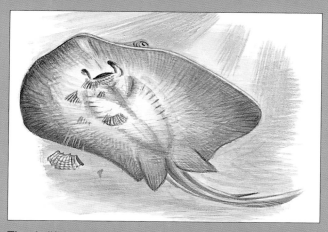

The shellfish is crushed in the ray's jaws. The inedible shell is spat out while the nutritious flesh is swallowed.

Index